4/03

also by phil patton

Dreamland:
Travels Inside the Secret World of Roswell and Area 51

Made in the USA:
The Secret Histories of the Things That Made America

Open Road:
A Celebration of the American Highway

Razzle-Dazzle

Voyager
(WITH JEANA YEAGER AND DICK RUTAN)

Highway:
America's Endless Dream
(WITH BERND POLSTER AND JEFF BROUWS)

bug

THE STRANGE MUTATIONS OF THE WORLD'S MOST

FAMOUS AUTOMOBILE

phil patton

simon & schuster

new york • london • toronto • sydney • singapore

SIMON & SCHUSTER
Rockefeller Center
1230 Avenue of the Americas
New York, NY 10020

For information about special discounts for bulk purchases, please contact Simon & Schuster Special Sales at 1-800-456-6798 or business@simonandschuster.com

DESIGNED AND COMPOSED BY KEVIN HANEK

SET IN ITC MENDOZA

Manufactured in the United States of America

10   9   8   7   6   5   4   3   2   1

LIBRARY OF CONGRESS CATALOGING-IN-PUBLICATION DATA

Patton, Phil.

  Bug : the strange mutations of the world's most famous automobile / Phil Patton.

    p.   cm.

Includes bibliographical references and index.

  1. Volkswagen Beetle automobile. I. Title

TL215.V6 P38 2002   629.222'2—dc21   2002070832

ISBN 0-7432-0242-2

bug

a t the end of the twentieth century the largest city on the planet is Mexico City, an immense sprawl of some 30 million inhabitants. From the air, glancing through the yellow-gray haze punctuated by two volcanoes, one sees not people but seven or eight million mobile metallic objects—automobiles. Coming closer, one can even distinguish the dominance of a single model, a buglike shape in a vast variety of colors, the majority green. These are the famous Volkswagen Beetle taxi cabs of Mexico City. Numbers and letters are painted on their roofs so they can be tracked by helicopter, but together, the characters seem to spell out some complex, secret message for the aerial visitor.

There are more Beetles here than anywhere in the world—nearly two million. The Vocho, Vochito, the Mexicans call the car, or "the navel," because everyone has one. In the thin air at 7,500 feet, their motors run at half power and despite fuel injection and unleaded fuel, many of them emit twice the emissions of modern cars. Reconfigured as taxis, with their front passenger seats removed, they run almost twenty-four hours a day.

Warned against the dangers of robbery in these taxis, most American tourists avoid them, and few of the tourists know that the original Beetles are still being manufactured in Volkswagen's factory a hundred miles from the capital. The 1998 high-tech, high-style New Beetle, a homage to the original, is rarely spotted in Mexico City's traffic and few owners of the car know that it is built, not in Germany, but only in Mexico, almost side by side with the old car.

Once tens of thousands of Beetles were produced in Germany, Brazil, Australia, Nigeria, and other spots around the globe. More than 22 million have been made since 1941. Mexico City is like a tidepool where the species has lived on long after the great wave of its success has receded. In this weird ecological microsystem the Bug survives, brand-new ones alongside ancient ones, in varying degrees of repair and in rich colors: wine, handpainted house white, lustrous silver blue, mineral green—and a rust brown rich as chocolate.

The Bug is not only the most produced, but the best-known car of all time. Its history is usually told as a glossed and gilded tale, glowing with nostalgia. But as in the story of Poe's gold bug, the insect itself is only the inspiration. The physical bug points the way to another, shadowy image of a beetle and a code, a secret language of wider cultural meanings.

The Bug's mental life far exceeds its metal one. It played many roles: poster logo for Hitler's National Socialist ideology, symbol of economic miracle in the Cold War years, paradigm of no-nonsense utility in the face of Detroit excess, icon of 1960s do-your-own-thing individuality, and—in the New Beetle—chic embodiment of fin de siècle retro.

The Bug's biography is a story of the way history can be forgotten, reshaped, and revived.

The Bug was contagious; it spread all over the globe and has kin in many countries: in Italy, the Fiat Topolino, or Little Mouse, in France the Citroën 2CV, in the U.K., the Mini, each with its own national characteristics. Even the Jeep, named for a Popeye character, has many of the same qualities the Bug acquired. More recent vehicles such as Chrysler Neon, Ford Ka, Renault Twingo, and Fiat Brava have attempted to build Beetle-like affection among the public. Each presents itself as the true heir to the Bug.

The Bug's direct descendant, the New Beetle of 1998, exaggerates the shapes and colors of the original like a cartoon. Although built only a few feet apart and sharing a rough common shape, the old and new Beetles are culturally as well as mechanically very different. The old has an air-cooled rear engine, the new a water-cooled front one. The old is driven by the rear wheels, the new by the front.

The original Beetle was a universal product of minimal ability, built to be cheap above all. The New Beetle is an object of style and pop culture. What the two share is a shape, a gestalt, a logos, as simple and winning as a cartoon or a popular tune.

The old and new cars share a childlike appeal. The re-creation of the old car's shape in the New Beetle brought with it the revival of a 1960s children's game. "Punch buggy" was one of those pieces of children's folklore, a game that called for a child who spotted and "called" a Beetle to strike another, exempt from retaliation. "Buggy" suggests either a horse-drawn vehicle or a baby's pram and even adults often colloquially refer to both old and new Beetles as "punch buggies."

Within a couple of years of being introduced the New Beetle became an icon of hip prosperity, a star in one of the Austin Powers films, which parodied the James Bond vision of 1960s gadgetry. It was the top prize in numerous sweepstakes, a familiar prop in advertisements and fashion shoots. It was inevitably included along with the Apple iMac computer in time capsules of the artifacts of the late twentieth century.

The two models provide an object lesson in the way images and ideas mutate through culture. They form a half-century-long parable of form and function, shape and association, a meme greater than the people who built and designed them, a complex of ideas and concepts and affections. The story involves a key paradox: how an impersonal universal design becomes an object of such personal attachment.

The Bug's history is not just the story of a single model of car, or even of the automobile in general, one of the two or three most profoundly influential pieces of technology to reshape daily life, but a parable of how the things we buy reflect the character of the culture. The Bug stands as proof that images and ideas swing through culture as if by their own power, evolving, adapting to new environments, latching on to new human champions, infecting human beings with enthusiasm. In some places and at some times, the Bug wore a self-deprecating mode of the servant, in others the personality of the émigré, the visitor, the friend and adopted guest.

Part of the universality and persistence of its appeal was the car's harmony between the two sides of design—the functional and the aesthetic. Its engineering and styling shared a common modesty and cleverness. Even the sound of its engine and the sensations of its movements on the road contributed to this. It crystallized the idea of the universal design with a human face. The Bug was cute and lovable, and soon electronic devices, appliances, even kitchen tools, aspired to be cute, cartoonlike friends: the smile on the Macintosh screen, the elfin character of the Kodak Brownie. The IBM standard PC "clone" was once described as "the VW Beetle of computers" and Steve Jobs pushed the Apple Macintosh as the "volkscomputer." A mouse looks more like a beetle than it does a rodent.

The Bug was a shape, a set of ideas—and a selfish meme.

The term meme is the creation of Richard Dawkins. In his 1976 book *The Selfish Gene,* Dawkins applied Darwinian theory in a fresh and startling way, arguing that not species, not individual specimens, but genes were the real actors on the stage of life. The individuals that carried the genes were merely vehicles whose purpose was to sustain the code for the species. "We are survival machines," he wrote, "robot vehicles blindly programmed to preserve the selfish molecules known as genes." Almost as an afterthought, he introduced the idea of the meme—the cultural equivalent of a gene.

In addition to genetic inheritance, wrote Dawkins, human beings have an adaptive mechanism that other species don't have: they can pass their ideas from one generation to the next. These memes, "ideas, songs, stories, images, concepts that passed from one generation to another, moving through culture and time," functioned like genes. Dawkins's meme sounded a lot like the Platonic eidos, the Jungian archetype, or the

German Gestalt. The difference was that in Dawkins's theory, memes transcended individual minds the same way genes transcended biological individuals. Memes, like genes, were "selfish."

But this suggested that a meme had human emotions—as if it were a character with a personality.

Dawkins provided an almost perversely fresh perspective. (No wonder he said that his book resembled "science fiction.") The selfish meme invited startling new ways to look at the questions: how and why do ideas and images and shapes persist through time?

But memes could not be charted like the genes of a fruitfly or human being. The meanings and identities of a meme would change in the way words change, taking on new meanings in new contexts, shifting from high speech to slang and back again. The Beetle was a meme that mutated this way, a shape and a set of ideas that fit Dawkins's description of selfishness.

The names Beetle and Bug for the car sprang up spontaneously not just in German but in every language—Käfer in German, Coccinelle in French, Fusca in Portuguese, Vocho in Spanish—inspired not only by the car's shape and buzzing engine but by its insectlike ubiquitousness. Bug, a common slang term for anything abundant and expendable—easily squashed—carried an air of informality befitting the car's character.

The word bug has a rich history and aura of meanings. Today a bug is most often used to mean something in electronics—a goof in the circuitry or any kind of small electronic device, most often one for surveillance, but it also refers to any human obsession, a fascination that grips a person like a disease, e.g., "He's got the bug."

At the beginning of the nineteenth century in England, bug meant bedbug and was not used in polite company. It flourished in the U.S. in slang—there were "tariff bugs" for instance—even before the Gold Rush of 1849 brought "gold bug."

Edward Tenner in *Why Things Bite Back* notes that the term bug was established by the middle of the nineteenth century as a term for an error or problem in a technical system. It seems to have been a familiar term in the world of telegraphers, when Thomas Edison began. A particular kind of telegrapher's key was already known as a bug in the 1850s, according to H. L. Mencken. In 1878, describing his method of invention, Edison used the word bugs to refer to all the "little faults and difficulties" that emerge to dog the inventor as he refines an idea.

The most famous application of the term computer bug came in the late 1940s. The chief programmer for the Mark II computer was Admiral Grace Hopper, whose name could have come out of a Disney film. One day after the computer failed, Hopper located a physical bug, a moth, near one of the computer's relays. She preserved it by taping it to a note card. With the arrival of personal computers, "software bug" entered popular language.

"It's not a bug, it's a feature" is an established joke of computer programmers facetiously justifying some eccentricity in software.

But isn't that the nature of the evolutionary mutation: that a bug turns out to be a feature, a mistake that benefits? In many ways the Beetle began as an aberration that turned into a triumph. First, it's a bug, then it's a feature.

While we usually think beloved, iconic products grow integrally from the cultures that produce them, the Beetle shows how shifting the culture around a product can change its meanings—and how a product can alter a culture.

It is an assumption of most cultural history that artifacts crystallize out of societies as neatly as rock candy out of a warm glass of sugar water, and that they are as characteristic and specific as Clovis spear points, Louis Quatorze chairs, Neapolitan espresso makers, hot dogs, Fabergé eggs, longbows, tail fin Cadillacs, and transistor radios. But the Bug suggests that designs, images, and ideas do not remain identified with the cultures that create them. They change as they move from one culture to another and alter the new environments that have altered them. Rolling through history on wheels of irony, the Bug lives a life somewhere between Woody Allen's Zelig and little Oskar, the protagonist of Günter Grass's novel *The Tin Drum*, and like them, it is a character that has managed to find itself at the center of key moments in history, including the very darkest.

chapter two

a dolf Hitler made his first formal appearance as Chancellor in a speech before the Berlin auto show. A centerpiece of Nazism was *Motorisierung*—the motorization of Germany—and on February 11, 1933, before an audience of automobile executives, Hitler laid out his plan to achieve this, a main tenet of which was to produce an affordable German automobile.

"There can be no triumph of National Socialism without radio, sound films, and motor cars," he intoned. "Simple, reliable economical transportation is needed. We must have a real car for the German people." He described a mass-produced car that could be bought by anyone who could afford a motorcycle. His vision for a people's car was of an automobile that would hold a family of four and cost less than a thousand marks—a price far lower than that for any car on the market at the time.

He declared an immediate end to the excise tax on automobiles and offered government support for motor racing. The German auto industry had produced only 42,000 cars in 1932, which was one car for every 100 people, compared to France, with one for every twenty-eight, and the U.S., with one for every six. He promised to increase Germany's output and to build new super roads for them to travel, which would improve the lot of the ordinary man—and raise Germany's prestige.

"A nation is no longer distinguished by the length of its railroads but by the length of its highways," he declared.

In Germany, which had given the world the internal combustion engine and its first applications in an automobile, the phrase "people's car" had been heard for years. Automobile clubs and magazines like the popular *Motorfabrik* and its editor, Joseph Ganz, crusaded for such a car; volkswagen and volksauto were already practically generic terms. A number of manufacturers offered models of cars advertised under these names, including Ford, whose "Cologne" model was produced in the factory it had opened in that city in 1931, and a small company called Standard. But support for a universal automobile was not just a fad Hitler picked up on, a whim or a propaganda gimmick; it was a vital part of his vision.

His fascination with the automobile and with the concept of a people's car in particular went back at least to the early 1920s. Henry Ford's autobiography *My Life and Work* had become a best-seller in Germany, and many people including Hitler believed that Ford represented a modernity and populism that could alleviate the suffering of the postwar years. A reporter who visited the Nazi party leader in 1922 found a huge picture of Ford on the wall in his office and piles of Ford's books on a table.

Hitler also believed the automaker shared much of his ideology. Ford's attacks on "financiers" and bankers in his book *The Great Present and the Greater Future,* published in 1927, seemed to feed into Nazism. Ford sponsored a periodical called the *Dearborn Independent,* whose pieces, collected in a book called *The International Jew,* linked Jews to alleged financial manipulations of worldwide banking and blamed them for the economic hard times. The Nazis had it translated into many languages and it became a best-seller in Germany. The leader of the Nazi youth movement, Baldur von Schirach, whose mother was an American, declared that he had become an anti-Semite after reading the book during his teen years.

Hitler also read motoring magazines, followed racing, and thought himself something of an expert on automotive engineering. This fascination with automobiles was one of the little bits of tinfoil his magpie mind had assembled among his historical, economic, and racist doctrines.

In 1923 after his unsuccessful putsch in Munich, Hitler was convicted of treason and sent to Landsberg prison. Sentenced to five years, he served only nine months, under conditions no worse than house arrest, and garnered wide public sympathy. It was one of the few times he read seriously in his life. He also dictated *Mein Kampf*.

Paul Johnson in his book *Modern Times* recounts that Hitler "spent his last weeks in jail thinking out the concept of spectacular scenic roads built specially for cars, the future autobahnen, and of the People's car or Volkswagen."

When he was released in December of 1924, he found a red Mercedes waiting for him, courtesy of the Mercedes representative in Munich, Jacob Werlin. Werlin would become Hitler's special deputy for automobiles and was vital in the birth of the Bug.

Hitler never learned to drive. The Führer was no *Fahrer*. When he wanted to show off his new Mercedes to Geli Raubal, the niece whose mysterious death Hitler covered up, he had to ask his chauffeur to drive him. But Hitler loved to pass other cars on the road, especially American ones, and delighted, he said, in riding high up into the mountains "where the goats gamboled around my great Mercedes." Standing in an open car became a signature Hitler pose. By 1935 Mercedes showrooms were already displaying giant portraits of him—behind the steering wheel.

For all his love of his luxury car, it was Ford's creation of inexpensive, widely available models that captured Hitler's imagination. By the 1930s the people's car already had a long history. It had come closest to realization in the U.S., of course. At the end of the 1920s, the U.S. had about one car for every seven people; half of all rural families had vehicles. The Ford Model T, introduced in 1908, was the best known and most successful people's car.

The American fascination with owning mechanical transportation had been evident in the bicycle craze of the 1890s, which helped lead to improved roads and brought many of the social changes in mores and recreation that the automobile would build on.

The private car was a welcome development as the public grew resentful of the barons' and magnates' control of the railroad. The automobile seemed a democratic and individualist alternative, and also offered freedom from fixed routes and schedules.

But like yachts or private railroad cars, the first automobiles were affordable only to the wealthy. Although the Duryea brothers had produced the first American cars in 1893, most were imported from Europe and showed up in Newport, Saratoga, and the estates of Long Island. A typical early car buff was William Kissam Vanderbilt, Jr. A millionaire dandy and the grandson of the famed Commodore, whose ships had built the family fortune, Willie Vanderbilt joined European aristocrats in races from Monte Carlo to Paris. In 1899 he came home from Europe with a new Mercedes, which he nicknamed the Red Devil. His terrified neighbors in fashionable Newport, Rhode Island, petitioned the police to impose one of the first speed limits in the country. In 1904 Vanderbilt set an automobile speed record for the mile, dashing at 92 mph along Ormond Beach in Florida. Soon he built his own highway. Beginning in 1906 he and the young heirs of the Belmonts and Whitneys played with their cars on Vanderbilt's concrete superhighway, the Long Island Motor Parkway, the first modern divided highway in America, which ran from Queens to Lake Ronkonkoma.

In 1904 he launched his own race, whose prize, the Vanderbilt Cup, was a ten-and-a-half-gallon silver creation fabricated by Tiffany. Over the years, the course of the road race varied, but part of it always ran on Vanderbilt's Long Island Motor Parkway.

In 1906, when there were just over 100,000 automobiles in the United States, Woodrow Wilson condemned the imperiousness of auto owners and expressed the fear that the car was polarizing classes. "Automobilists are a picture of arrogance and wealth," he declared. "Nothing has spread socialist feeling in this country more than the automobile."

Two years later, Henry Ford introduced the Model T.

The key to a successful business, Ford was convinced, was to find one thing that everybody wanted and make that and nothing else. Early in his career he considered mass-producing inexpensive watches, but he calculated that he would have to sell 600,000 watches to break even. The automobile, by contrast, offered a virgin market. Behind Ford, too, stood American industry's history of assembly from interchangeable parts of firearms, sewing machines, typewriters, and bicycles.

Ford would be lionized as the master of mass production, but his goal was never simply maximizing production. When he outlined the characteristics of his ideal motor vehicle in *My Life and Work,* which Hitler read in Landsberg prison, surprisingly few of them had to do with mass production. They included lightness, simplicity, durability, and ease of repair. Parts should be simple to replace and universally available. Above all Ford aimed to build a "universal car"—a single model for all. "Ford The Universal Car" was the motto on the winged symbol displayed at garages and dealers in the 1920s and 1930s.

A "universal car" meant a car "for the great multitude, constructed of the best materials after the simplest designs that modern engineering can devise, so low in price that no man making a good salary will be unable to afford one." There was a social benefit to such a car. With it, Ford said, the working man would be able to flee the city and take his family into God's great outdoors.

This idea of one simple model for everyone could be applied to other products. Ford believed that "the less complex an article, the easier it is to make, the cheaper it may be sold, and therefore the greater number may be sold." Standardization, he explained, is the final stage of the process that begins with a consumer's needs, proceeds through design, and finally arrives in manufacture. But he warned of a tendency to keep monkeying with styles and to spoil a good thing by changing it.

"It is considered good manufacturing practice, and not bad ethics occasionally to change designs of the old models so they become obsolete . . . we cannot conceive how to serve the customer unless we make for him something that as far as we can provide will last forever . . . we want the man who buys one of our products never to have to buy another."

Ford's plan was to make all the parts on his cars interchangeable not just with those of the same model, but those of past models. Cadillac had pioneered interchangeable parts as early as 1906 for precision in performance and ease of repair. Ford's use of them would achieve dramatic economies in production.

His interchangeable parts promised to protect the buyer against obsolescence, but they limited future improvement. Here in a nutshell was the promise and paradox of the universal product of all types: low cost and simplicity came at the price of individuality and innovation.

"When I'm through," Ford said, "everybody will be able to afford one and about everyone will have one." Universal ownership of the automobile reshaped the economy, the society, and the landscape in the image of the car. The unforeseen result of a world where everyone could afford a car was that no one could afford to be without one.

This was one of the most significant social, cultural, and economic developments of the century: the eventual consequence of making the car such a necessity was that people were helpless without one.

Ford introduced the Model T in 1908. By 1914, when he had established the famed moving assembly line, the T's price had been cut to $550—more than a third less than it had been in 1908. The car's limits did not become clear until 1927, when sales fell by a third, but by then it had become part of American culture as well as daily life. The T became a character, ruefully admired and cussed: it was the "flivver," the Tin Lizzy, the name Lizzy from the common designation for a black female servant. It was a cheap but indestructible servant and sidekick—a steel Mammy, a tough Mama.

Lizzie was clearly kin to other mass-produced black multiples, like George Eastman's Brownie camera, introduced in 1900, triumphs of abbreviation and repetition, or mass-media characters like Chaplin's Little Tramp in his black bowler and Mickey Mouse, whom Walt Disney had described as embodying the American spirit of "the great unlicked." The T was a Little Tramp of the American Road cheerful in mud and adversity in its standard black suit. "Any color you want as long as it's black" was a wry spin on the reason black was chosen—for ease of manufacture. It was the only color that dried fast enough. The Model T was famous for the jokes told about it. A man sent a package of tin cans back to the Ford factory and in return received a new Model T, with the accompanying note, "Although your car was severely damaged we have managed to repair it." While the T had no doors, one joke went, can openers were available from the company. A T without oil or gasoline ran thirty miles on its reputation. A Model T ran over a chicken, which emerged unscathed saying "cheap cheap cheap."

In his famous essay "Farewell My Lovely," E. B. White described the T's upright windshield and the high ground clearance that made it successful in rural areas with rutted roads. The T was "hard-working, commonplace, heroic," and White noted that "it often seemed to transmit those qualities to the persons who rode in it." The driver of a Model T, he wrote, was "a man enthroned."

In the early 1960s, designer Jay Doblin praised the T as the most purely functional automobile ever designed, and for that reason alone, he argued, "A strong case can be made that the Model T is the most beautiful car ever built." But the T also had its critics. One charged that while "it is a byword for utility and efficiency to the unthinking . . . the lack of rhythmic balance in its organization, its stupid, sterile vertical lines, frustrate all consciousness or horizontal motion and velocity."

Hitler's speeches at the Berlin auto show became an annual ritual. Automakers had done little to realize the vision for the people's car that he had articulated in 1933 and when he returned the next year, he called on them once more to manufacture a people's car. And he again struck the populist note, subtly nurturing class resentments.

"So long as the motor car remains only a means of transportation for especially privileged circles," he said, "it is with bitter feelings that we see millions of honest, hard-working, and capable fellow men whose opportunities in life are already limited, cut off from the use of a vehicle." Owning a car, the Führer said, "would be a special source of yet unknown happiness to them, particularly on Sundays and holidays." He declared confidently that the people's car would soon arrive on the market and used the term volkswagen for the first time. What he did not say was that he believed he had found the man to build him his people's car.

chapter three

during the 1920s, Hitler enjoyed going to the auto races run at the AVUS track in Berlin. Completed in 1921, it was made up of two two-lane stretches of pavement with a median and ran for six miles from Berlin toward Potsdam through the Grunewald Park. AVUS was an abbreviation for Automobil Verkehrs- und Übungs-Strasse, which translates roughly to Automobile Traffic and Practice Street. Built as a racing circuit, the track was also intended to make up the first portion of a national highway system.

At the German Grand Prix of 1926, Hitler went up to a disheveled man in a trench coat standing beside the track, oblivious to the noise and smoke and speeding cars. He was a strange figure somewhere between a football coach and mad scientist, but fans would come up to him after races to shake his hand and perhaps get an autograph.

There was no reason the perpetually preoccupied Ferdinand Porsche should have remembered the local politician, despite the man's wild shock of hair and mustache and his excitement at meeting the great engineer. But Adolf Hitler remembered Porsche.

Porsche and Hitler, each in his own way, felt some affinity, however idealized and self-deluding, with the "common people" from whom they had sprung. Both men also cultivated a resentment of aristocrats and corporate executives, whom they viewed as men of selfish and limited vision.

Like Hitler, Porsche came from the edges of the decaying Austro-Hungarian Empire. He had been born fourteen years before Hitler in the village of Maffersdorf, Bohemia, and, much like Hitler, grew up in the provinces before moving to Vienna.

Porsche never lost his sense of the village and region he came from. Porsche owned a series of country retreats and hunting lodges throughout his life, and he outfitted cars as hunting vehicles with storage compartments for gear and game. His wife, Aloisia, an ambitious young bookkeeper, was also from Bohemia and similarly enjoyed ties to a rural life. Their first daughter, Louise, was born there in 1904, and their first son, Ferry, who would become his father's closest business associate, in 1909.

The child of a tinsmith, Porsche was the classic precocious tinkerer. While still in his teens, he installed electrical wiring in his family's house. He went to work for electrical companies in Vienna, but was soon drawn to automobiles. In 1902 he became nationally known as an engineering prodigy when he developed a mixed diesel and electric power system for the Lohner company, a small automotive firm. The Mixte system was a forerunner of hybrid engines of eighty years later. Porsche once demonstrated this system to Archduke Ferdinand.

At the beginning of the First World War, which was ignited by the Archduke's assassination, Porsche was with Austro-Daimler, which was soon absorbed into the larger firm of Škoda, and Porsche's mixed drive was put to work moving huge artillery pieces.

In the 1920s, Porsche designed great Mercedes racing and road cars. He was acknowledged to be a brilliant engineer, but his terrible temper made him hard to work with. In his fits of rage, he would often throw down his hat and stomp on it like some silent film character. The idea of a small inexpensive car for people like those he had grown up with seems to have been with Porsche from the beginning of his career. As early as 1921, when he was working with Austro-Daimler, he unsuccessfully pushed the company's management to develop and sell a smaller, less expensive model. In 1928, after he moved to German Daimler he again pushed for a small car, without result.

In 1930, frustrated with working for larger companies, Porsche established his own independent engineering consultancy. To speak of Porsche from this point on is to speak of his team: Karl Rabe, perhaps his most important lieutenant and an engineer whose brilliance Porsche quickly recognized; Joseph Kales, the engine man; Hans Mikl, the aerodynamicist; Franz Reimspiess, motors and general engineering; and Erwin Komenda, bodies. Adolf Rosenberger was the business head, but a Jew, he left Germany just before Hitler took power and ended up having a successful career in the automotive community in California, where he took the name Alan Roberts.

To get his company going, Porsche drew up sketches for a novel new race car, designed a luxury car for the firm Wanderer, and tried to find a builder for his people's car. But times were hard and most of the dozens of auto firms in Germany were small and on the edge of bankruptcy.

In 1931, Porsche learned that Zündapp, a popular motorcycle maker, was ambitious to add a small car to its line. Porsche offered a design using his new and inexpensive torsion bar rear suspension. One of his great innovations, the torsion bar replaced traditional coiled springs with a tube filled with metal rods whose twisting served the springing function. It became a basic part of automobile construction around the world, and royalties from the design helped support Porsche's independent engineering firm.

Zündapp's owner, Fritz Neumeyer, was willing to try Porsche's torsion bar suspension but not the engine he suggested. Neumeyer insisted on a heavy radial five-cylinder water-cooled motor. When the prototype was tested, the engine overheated, the car began shedding gears and other parts, and Porsche's torsion bars—made of inferior steel—shattered. The car and the project ground to an ignominious halt. Zündapp decided to stick with motorcycles.

Still, the rear-engine Zündapp—Type 12 in the sequential numbering system the Porsche company gave its projects, but called the Volksauto by Neumeyer—was a rough sketch of the ideas that would show up in the Bug.

A more finished draft of Porsche's people's car concept was the Type 32, built for NSU, another struggling company. NSU's boss, Fritz von

Valkenhayn, also wanted to add a small car to his product line. He let Porsche try an air-cooled rear engine and a body whose shape foreshadowed the Bug's. But in 1933, after he had built three prototypes, Porsche was informed that NSU had abandoned the project and had formed an alliance with Fiat.

In 1932 Porsche was invited by top party leaders to the Soviet Union. The Russians were looking for a man to design and build them a sturdy and versatile tractor, and he was wined and dined and taken on a technical tour of the best of Soviet factories and workshops from Kiev to Kursk to Odessa. He even saw secret industrial centers behind the Ural Mountains.

He was offered the job of official state designer, with thousands under his command, a blank check, and a villa of his own. He would create not just a tractor or a car but a whole modern automotive industry. There was one restriction, however: he would not be able to leave the country. Soviet ideology did not bother him, but he could not imagine himself at home in the U.S.S.R. He needed family and gemütlichkeit. The vast spaces of the Russian landscape (the spaces Hitler later planned to fill with German villages, German autobahns, and German cars), one source records, "seemed strange to him and somehow sinister." He turned down the offer and returned to Germany.

With new resolve, he focused his company on three goals: to create a race car that pushed the edges of technology, to create a people's car, and to create a tractor to improve rural life.

Germany had dozens of small auto firms, with strange names like Presto and Steyr, but in the Depression most of them lived on the edge of bankruptcy. On the heels of Hitler's 1933 speech to the Berlin auto show, they began angling for whatever government support was available. One of the savviest was Auto Union, Germany's second largest automaker, formed in 1931 from four Depression-strapped firms, Wanderer, Horch, DKW, and Audi. The company logo was four linked rings symbolizing the four companies. In May 1933, the head of the Auto Union board, a wealthy baron named Klaus D. von Oertzen, managed to obtain an audience with Hitler.

The regime had offered up 500,000 marks (a quarter million dollars) to support racing, but had quickly directed the entire subsidy to Daimler-Benz, the country's leading carmaker—and Hitler's favorite—and the one expected to do the best against international competition. To change the Führer's mind, von Oertzen brought two calling cards: Hans Stuck, the race driver, and Ferdinand Porsche, who had designed Mercedes's top race cars in the mid-1920s and was now working for Auto Union.

In the 1920s, Stuck had lost his job racing cars for Mercedes, but he had a friend named Julius Schreck, a chauffeur for the local politician Adolf Hitler. After Stuck was fired, he asked Hitler for help, and Hitler promised him that when he came to power, Stuck would get a car to race.

Now von Oertzen and Stuck argued that Auto Union should receive part of the government subsidy because it had a plan for a new and better kind of race car. The man behind this plan was Ferdinand Porsche. "We've met before," Hitler said to Porsche, recalling the 1926 German Grand Prix.

Then for twenty-five minutes, without brooking any interruption from Hitler, Porsche laid out his plans for his racer. It was a design he had had on the drawing board for a year. Very light—lighter, it turned out, than the future Bug—it would be powered by a V-16 engine generating nearly 600 horsepower and placed over the rear wheels for better traction. But while it could reach nearly 250 miles per hour, it would also turn out to be notoriously difficult to drive. This design for the Auto Union racing cars anticipated the Bug in such features as torsion bar suspension, engine placement, and even in the bodywork of Erwin Komenda.

Persuaded by Porsche's presentation, Hitler decided to split the subsidy between Auto Union and Mercedes. Porsche used his usual team: Joseph Kales did the engine, Karl Rabe designed the suspension, Komenda shaped the body, and the Auto Union car was running before its competitor.

At the AVUS, Hans Stuck test-drove the car, the so-called P-Wagen, for the first time in January 1934, topping 155 miles per hour, then raced it that May. It ran well until the clutch broke. But back at the AVUS for the German Grand Prix later that year, Stuck outdueled the Mercedes driven by Rudi Caracciola to claim victory. Stuck went on to win the

Swiss Grand Prix by a full lap and drove the P-Wagen to the European championship.

Thus began the era of the Silver Arrows, when race cars shaped like fighter plane fuselages dueled in front of fevered crowds. According to legend, the Mercedes entry at the Eifelrennen race at the Nürburgring in 1934 was found to be a kilogram overweight. Its supporting crew spent all night sanding off the paint, and the next day the raw aluminum racer slipped under the weight limit. Soon both manufacturers raced their increasingly streamlined vehicles in bare metal or silver paint. Silver replaced white as the racing color of Germany.

The races presented the Nazis with a tremendous propaganda opportunity. They offered speed and spectacle, technology and competition, and symbolized a revived German culture and economy. The races supplied hours of radio programming and exciting newsreels and newspaper photographs. The stands were filled with the spectacle of Nazi banners and uniforms. In the 1930s the Nazis gave the AVUS track a steep brick turn where cars could be photographed to dramatic advantage. A round tower was built with three balconies, offering party officials and guests a perfect view of this soon-to-be-famous "Nordturn." More than his call for a people's car, more than the autobahns, the races were the most visible and dramatic manifestation of Hitler's motorization program.

The research that went into race car engines helped develop aircraft engines, and the races themselves served as prewar pep rallies: at one race, Stuka dive-bombers, which would become the very symbol of the terror of the Blitzkrieg, put on a demonstration for the crowds, plunging shrieking toward the track.

Hitler controlled the sport under the National Sporting Commission for German Racing, or Oberste Nationale Sportbehörde für die Deutsche Kraftfahrt. All drivers were forced to join the National Socialist Driver Corps or Nationalsozialistisches Kraftfahrer Korps. To run the commission, Hitler appointed a World War I veteran, Korpsführer Adolf Huhnlein, a humorless martinet given to the most elaborate Nazi regalia—a buck-toothed Nazi as caricatured by Mel Brooks, a Nazi whose self-importance was so overblown that the drivers and even other Nazi officials mocked him. At one victory ceremony, the winner was presented with a bust symbolizing the Goddess of Victory. While Huhnlein's back was turned, the star driver Berndt Rosemeyer stuck a burning cigarette in the Goddess's mouth.

The AVUS was a temple not only to auto sport but to the national fascination with modern superhighways, which auto associations and publications had been advocating for decades. It had been built to show what a divided highway would be like and, except when races were run, it was open to ordinary traffic.

Its proponents were one of many groups pushing a national autobahn plan, and Hitler made their cause his own. When Hitler came to power, one of his early followers, a civil engineer and honorary SS colonel named Fritz Todt, already had an autobahn plan. Written in December 1932, and titled "Proposals and Financial Plans for the Employment of One Million Men," it called for building some 5,000 kilometers (around 3,000 miles) of autobahns. The task, Todt calculated, would employ 10 percent of Germany's unemployed. From the Nazis' standpoint it also had the advantage of bypassing the private sector.

By the end of June 1933, Todt was installed as inspector general of the German road and highway system. Hitler himself turned the first spade of sod in September 1933, and was able to cut the ribbon on the first stretch of autobahn, from Frankfurt to Darmstadt, by the fall of 1935. Six hundred miles of roads were built in three years, at a rate of about six-tenths of a mile a day. Todt, whose name sounds like *Tod*, the German word for death, understood modern highway engineering and expounded a theory of the "philosophy of the beauty of the highway," which held that gas stations and other buildings should blend with the local landscapes and bridges should use local materials.

Foreign visitors were invited. British professor F.G.H. Clements viewed the roads from the Graf Zeppelin and returned home raving about "the sheer clean beauty and vigorous sweeping curves of the autobahns." An anonymous writer for the *American Magazine* in 1936 praised the "simplicity and crispness of design" of the bridges and other features. Each road, he reported, was made up of

> two roomy concrete lanes in each direction separated by a 16 foot grassy median. . . . To either side considerable strips of land are preserved as integral parts of the scheme, preventing access to the roadway and possessing definite aesthetic advantages. All curves are wide and well saucered for high speed; on the Frankfurt–Darmstadt stretch, I rounded a curve at over 60 mph without touching the steering wheel, the set of the car's gravity in the bank carrying it around and straightening out the wheels automatically with the road.

Stephen Roberts, a history professor at the University of Sydney in Australia, visited Germany from November 1935 to March 1937. So unfamiliar were English, Australian, and American readers with the concept of the divided highway that Roberts felt obliged to describe the new method of driving the autobahns required. "Travel on the new roads is rather peculiar," he explained in his book *The House That Hitler Built.* "With no fear of cross traffic and with a perfect concrete surface there's no speed limit. While being officially conducted over these roads I have taken notes when driven at 80 or 90 miles an hour."

The autobahns impressed even Viktor Klemperer, the eloquent Jewish diarist of the Hitler years. In 1938 he recorded:

> Half by chance we found ourselves on the new Reich autobahn from Wilsdorf to Dresden, less than an hour after it was opened. There were still flags and flowers from the ceremony in the morning, a mass of cars moved slowly forward at a sightseeing pace. Only occasionally did anyone attempt a greater speed. This straight road, consisting of four broad lanes, each direction separated by a strip of grass, is magnificent. And bridges for people to cross over. Spectators crowded onto these bridges in the sides of the road. A procession. And the glorious view as we were driving straight toward the Elbe and the Lossnitz Hills in the evening's sun. We drove the whole stretch and back again. . . . Twice I risked a speed of 50 mph.

Most of Germany's roads were in terrible shape. "There are constantly accidents at hundreds of railway level crossings, thousands of roads are in the worst condition." It made more sense, Klemperer was sure (and as the transportation bureaucrats had told Hitler in the spring of 1933), to fix them before building new ones. "None of it is done because of course it would not catch the eye: the ROADS of the Führer!" But as economic stimulus and national morale builder, the autobahns seem to have worked at least as well as American New Deal programs such as the Civilian Conservation Corps and Works Progress Administration, which put the unemployed to work building highways, parks, post offices, and dams. In *The Nazi Economic Recovery, 1932–1938*, the British economic historian R. J. Overy argued that motorizing Germany did more than rearmament to pull the economy out of its doldrums.

The automobile industry tripled production in 1933 and road building provided a stimulus for key sectors of the economy. Cars and roads were linked to important modern industries such as steel, coal, and concrete. Road building was semiskilled work that absorbed unemployed workers. Although the number of unemployed put to work on the autobahns never matched Todt's promises, those who did find work were no longer being supported by the state, saving 120 million marks, the Nazis bragged. The project also helped the Nazis control workers who might have opposed them. They made work on such projects compulsory for the older unemployed and established them in a semimilitary condition. They lived in wooden barracks and ate from a common mess. For younger men, the chance to work on the autobahns dampened criticism from the regime's political opponents.

The higher speeds of the autobahns put a new emphasis on streamlined vehicles. By the mid-1930s Mercedes was producing sleek models called Autobahn Couriers. Stephen Roberts noted how often speed figured in Nazi propaganda about the autobahns. "Dr. Todt himself has bought a supercharged car and Hitler has stated that his example of traveling at a hundred miles an hour will gradually become the norm." "To survive in these years of trial," one Nazi official told Roberts, "German life must be speeded up to a hundred miles an hour."

Motorization was openly linked to militarization. The autobahns ran toward the borders where Germany planned to expand. The notions of "national renewal" and muscle flexing tied neatly to rearmament. Thomas McDonald, chief of America's office of highway administration, and officials from the Pennsylvania Turnpike Commission, touring in 1939, were impressed by the highways but more so by the columns of military equipment sweeping by them at high speeds.

By 1942 when the war brought an effective end to construction, some 2,326 miles of the autobahn system were complete. Hitler expanded plans to nearly 10,000 miles as his imperial designs grew. "We can no longer think of the German Reich without these roads," Hitler declared, adding ominously that "in the future they will also find their continuation as natural great communication lines in the whole European transportation region."

The autobahns cried out for more cars to run on them, and the people's car offered a highly effective propaganda tool. It played on class resentment and held up American-style prosperity as a goal. Hitler had called for both highways and automobiles, but the highways turned out to be easier to build than the cars.

Not long after Porsche met with Hitler to discuss the Auto Union race car, Jacob Werlin, the Mercedes representative who had become Hitler's key auto adviser, stopped by Porsche's offices in Stuttgart. Porsche at first suspected Werlin had come to pump him for details on the planned Auto Union racer, but Werlin had another agenda.

Hitler's call for a people's car had met with skepticism on the part of established automakers—including Mercedes. Most considered it unrealistic. But Porsche did not. Werlin had heard of his long-standing interest in a small car and experiments for Zündapp and NSU. Would he perhaps be interested in discussing these with the Führer? After Werlin's visit, Porsche hastened to prepare a plan.

In the fall of 1933, while the Nazis consolidated power, Porsche worked out ideas for the little car. His initial model was the earlier people's car he had designed for NSU.

By December 1933, it appears, Hitler was already making notes and sketches in consultation with Porsche and drawings, which he signed A.H. Porsche offered his by now familiar concept—a rear-mounted, air-cooled engine and torsion bar suspension. Hitler wanted an air-cooled car because it would start better in the cold and most Germans did not have garages. The car should be able to hold a family of five, he said, for trips to the country, and maintain a steady speed of 80 kilometers per hour, or between 40 and 50 miles per hour. They discussed costs. Porsche's carefully thought-out proposal included a planned price of 1,500 marks. Build it, Hitler told him, at any price—as long as it is under 1,000 marks.

In January 1934, Porsche officially submitted his "Exposé for a German People's Car." The accompanying sketches showed a vehicle very much like the earlier NSU prototype, with a narrower body and wider fenders than the eventual design and the headlights perched close together above the front hood. But the basics were there: rear air-cooled engine and torsion bar suspension.

The key to Porsche's proposal was an understanding that a new scale of car required new design. "It should not be a pantograph solution," he said of the people's car, meaning a shrunken version of a conventional car, as if reduced by a draftsman's pantograph, a device that traces a shape and reproduces it at a prescribed percentage of the original.

The car would carry a family of four, and even with an engine of only about 25 horsepower it would cruise at 60 miles per hour, climb a 30 degree grade, and get about 30 miles per gallon. It would weigh about 1,400 pounds and have a 1,250 cubic centimeter engine running at 3,500 revolutions per minute.

"I have thoroughly studied the question of a people's car," Porsche wrote in his proposal, "and I cannot consider it as a mini car which by artificial scaling down of dimensions, output, weight, etc. continues along the well-trodden paths of the traditional vehicle. It might be cheap to buy originally, but . . . its value in use would be small due to reduced driving comfort and durability.

"My interpretation of a people's car is a useful vehicle able to compete with any other car on the same level. If the traditional car is to be made into a people's car, then fundamentally new solutions will be necessary."

But to approach Hitler's 1,000 mark price limit, radical innovations would be required. To provide a durable ride at low cost, Porsche relied on his invention, the torsion bar. The front wheels used a trailing arm suspension, but in the rear, a torsion bar was combined with a swing axle suspension. This arrangement made for a softer ride by angling, or cambering, the wheels slightly. But this made the rear-wheel grip less predictable, and in combination with the higher proportion of the car's weight to the rear, helped give the car its characteristic tendency to oversteer.

The key to the car's success was its engine. Aiming at the lowest possible manufacturing cost, Porsche decided on an air-cooled, rear-mounted engine. An engine's mode of cooling is as fundamental to its character and efficiency as its mode of combustion. Of the energy produced by an engine, only a small fraction is captured as kinetic energy— the power that moves pistons and gear shafts and wheels. The rest is heat that must be transferred to the environment or the engine will overheat. The most common cooling method is to transfer the heat to water, which is pumped around the engine and then into a radiator, which in turn is cooled by a fan. But the resulting weight, as well as the power needed to run the pump and fan, put additional loads on the engine.

In contrast, in air-cooled engines, used for years in motorcycles and aircraft, the cooling fins and fan alone did the job. Air cooling not only eliminated weight but many parts, potential leaks, boil-overs, freeze-ups and other potential problems, and it made the car more resistant to extremes of temperature. It was cheaper, too. Porsche's genius was to adapt this system to low-cost, mass production.

Rear mounting of the engine eliminated the heavy and bulky drive shaft and put the weight of the engine over the driving wheels. Structurally the car had a central spine and a flat floor pan that could be stamped out. At first, Porsche considered using plywood for the floor.

The engine itself featured a central crankshaft, longitudinally placed. Initially Porsche envisioned a simple and economical two-cylinder motor, but none of the designs proved satisfactory. Four cylinders were then considered, but neither Joseph Kales, Porsche's regular motor expert, nor others in his team were able to produce an acceptably light and rugged model. Porsche even considered using a World War I aircraft engine.

The task of creating the right motor finally fell to Franz Xaver Reimspiess. The thirty-three-year-old Reimspiess, a confident man with an intense, intelligent face, his black hair slicked back to form a widow's peak, had caught Porsche's eye when he solved a mechanical problem that had frustrated the older engineers. Porsche recognized in him the same versatile and innovative talent he had seen years before in Hans Rabe, whom he had picked out at age eighteen and who became Porsche's master engineer and alter ego on the shop floor.

In just days, Reimspiess drew up sketches for a flat four engine, also called a boxer motor for its cylinders, which face each other like the fists of opposing pugilists. Reimspiess made the parts as light as possible, and shaped them for durability, not raw power. He also managed to solve problems of cooling the engine oil that had stumped Porsche's other engineers.

The transfer of heat was managed by a fan that blew air over a heat sink, an arrangement of thin metal plates or fins surrounding the engine block that radiated the heat into the air. The fins were also a beautiful design feature, suggesting a kind of attenuated Art Deco style, and made the solid motor seem to fade gradually into the space around it.

Reimspiess designed the engine to run long, not hard. It would operate in the center of its ability, not the extremes. It would spend its life doing what it could do best, not doing the best it could. Its buzzing sound, so easily recognizable, seemed to speak of persistence and beelike industry.

Millions of copies of Reimspiess's motor and its descendants have been turned out since 1938, with improvements and variations for use in land vehicles of all sorts, as well as in light aircraft and home electrical generators.

In January 1934, Hitler returned to the Berlin auto show and urged industry to produce "a true car for the German people, a Volkswagen."

Hitler's first call for a people's car had left German automakers unmoved. They had no idea how anyone could produce the car Hitler wanted—certainly not at a price under a thousand marks—the political benchmark. A few offered their own approximations of what a practical people's car would be like, coming as close as they could to what Hitler wanted. Mercedes showed its small, rear-engine Model 130 at the 1934 Berlin auto show. Leaving no doubt that they were trying to impress the Führer, the company offered a camouflaged military model side by side with the civilian one. Not until 1937 did Opel, the largest seller of cars in Germany and owned by General Motors, present its small P4, and at 1,430 marks its price was nearly half again as high as Hitler wanted. DKW, the small firm known for motorcycles and very cheap cars and one of the four firms bound into Auto Union, did not get its P9 ready until 1939. For the most part, the automakers hoped the people's car idea would fade away as the difficulties became clearer. They may have thought it was all propaganda, or they may simply have underestimated the grasp it held on Hitler's imagination.

During the 1934 Berlin show, Hitler met with Porsche at his suite in the Kaiserhof Hotel, and by May he was commenting on Porsche's ideas and sketching his own. The first drawing for what became Type 60 in Porsche's numbering seems to have been made on April 27 for discussions with Hitler on May 12. It was on that day that Hitler apparently made his famous statement that the car should look like a beetle.

The engineer and the dictator did not agree on every point. Porsche wanted to use hydraulic brakes instead of mechanical ones, but the patents were held by the British firm of Lockheed; Hitler wanted the car to be all German. As a result, brakes would be the weak point of the car until after World War II. Porsche also wanted the body and chassis to be more fully integrated, but Hitler insisted that the chassis accommodate different bodies for military purposes. The Reichsverband der Deutschen Automobilindustrie (RDA), comprising the leading automakers, including Daimler-Benz, Auto Union, BMW, Opel, and others, moved to respond to Hitler's pressure.

In June, the RDA signed a contract with Porsche, to develop prototypes for a people's car by the middle of 1935. The RDA would then evaluate the prototype and recommend further action. But the question of who would actually build the car remained open. The companies did not want any competition, especially from the government.

All were skeptical of Hitler's ideas and most on the RDA committee expected Porsche's effort to fail. They questioned the practicality of his technical ideas; Opel's representative on the RDA committee, Heinrich Nordhoff, derided the planned air-cooled engine as "an airplane engine" that could never be manufactured at acceptably low cost. Others thought Porsche was simply trying to get Nazi money to keep his business going.

Construction of the prototypes by Porsche's staff and the Stuttgart coach builder Reutter began in December, but the first car did not travel under its own power until October of the following year. The first prototype was shown to Hitler in December 1935, or January 1936, at his Obersalzburg retreat. The miffed officials of Porsche's nominal client, the RDA, were not shown the car until February, but Porsche understood who his real employer was.

Testing of the prototypes, called VW3s, began in the fall of 1936. Between October 12 and December 22 the three VW3 models were driven some 50,000 kilometers or around 30,000 miles, the exhausted test drivers rushing to complete their schedule before Christmas. The cars held up well, except for a problematic crankshaft that had to be strengthened. One team of drivers struck a deer. The car survived, and the venison became part of a holiday dinner.

The next year, Daimler-Benz was contracted to build, virtually by hand at a cost of about $50,000 each, thirty more mature prototypes, thus making some of the first Volkswagens actually Mercedeses. Several changes distinguished the new cars from the VW3s. The headlights were blended into the front fenders instead of mounted on the hood, a change that essentially established the familiar VW face. Two rounded panes of glass were mounted in the vented rear deck so the driver could see behind him. This second crop of test cars, called V30s, were driven by a corps of SS men—not professional drivers—sworn to secrecy and billeted in a tank barracks.

In the fall of 1937 each VW30 was driven nearly 80,000 miles along twisting roads in fairy-tale forests, on the new concrete autobahn between Stuttgart and Bad Nauheim, and through the Grossglockner Pass in the Alps. Altogether the cars accumulated an unprecedented 2.4 million miles. The results impressed even the RDA, which despite the skepticism of the member companies, gave the car a fair technical evaluation and reported it to be a quite durable vehicle.

chapter four

I t should look like a beetle," Hitler is said to have advised Porsche, as he sketched his ideal shape for the Volkswagen in May of 1934. "You need only observe nature to know how to achieve streamlining."*

The name would stick to the car. Käfer,† Beetle, Coccinelle.

The Führer had in mind a very specific beetle: the Maikäfer, the May beetle, *Melolontha melolontha,* a member of the family *Scarabaeidae.* The Maikäfer resembles the English cockchafer, relative of other brown, round beetles found around the world, buzzing around porch lamps in warm weather, their wing covers lifted like gullwing doors on a sports car. The cockchafer feeds on decaying vegetable matter. It is kin to the American June bug, the green-blue beetle to whose legs children tie strings, but it also figures in German culture like the spotted English ladybug: a child-friendly insect from nursery rhyme and fairy tale.

> Lady bug lady bug fly away home,
> Your house is on fire and your little ones gone.

* Hitler's original German: *"Wie ein Maikäfer soll er aussehen. Man braucht nur die Natur zu betrachten, um zu wissen, wie sie mit der Stromlinie fertig wird."*

† The restaurant at VW's Berlin Forum, on the Kurfürstendamm, is called the Käfer Café. The name comes from its owner—a man named Käfer, whose original restaurant, of the same name, is in Munich.

It is a relative of the Egyptian scarab, a creature, naturalist Sue Hubbell reports, associated with the Egyptian word for "to become." It is also kin to Albrecht Dürer's stag beetle, painted with an exacting observation of nature, and to the Mistkäfer, or dung beetle.

This is the insect into which Gregor Samsa, the hero of Franz Kafka's story "The Metamorphosis," finds himself transformed one morning when he wakes up. (According to Vladimir Nabokov, entomologist as well as novelist, it was not a cockroach but a dung beetle. The creature's back is rounded, Nabokov notes, not flat like a roach's, because he has trouble turning over on his bed and his legs wave in the air. He has strong enough jaws to turn a key in a lock and can stand upright—characteristics of a member of the scarab family, but not of a cockroach. The old charwoman in "Metamorphosis" even addresses him, perhaps facetiously, perhaps descriptively, as a Mistkäfer—dung beetle. "At first she even used to call him to her, with words which apparently she took to be friendly, such as: 'Come along, then, you old dung beetle!' or 'Look at the old dung beetle, then!'")

"You need only observe nature . . ." The VW did look like a beetle, with headlights as eyes and even wing shields on its hood. To strengthen the resemblance, Hitler drew a more rounded front for the Bug. Porsche was not slow to pick up on the suggestion. To look like a beetle—to look like fish, or bird, or water drop—to look to nature for streamlining was the most common oversimplified cliché of aerodynamics.

As Hitler delivered his advice to Porsche in 1934, Chrysler was making an almost identical argument in its advertisements for the streamlined Airflow. "Old mother nature has always designed her creatures for the function they are to perform. She has streamlined her fastest fish. . . . Her swiftest birds. Her fleetest animals that move on land. You have only to look at a dolphin, a gull, or a Greyhound to appreciate the rightness of the neighboring, flowing contour of the new Airflow Chrysler. By scientific experiment Chrysler engineers have simply verified and adapted a natural fundamental law." Chrysler's Airflow may have been aerodynamically advanced, but its appearance put off customers.

William Stout, who helped develop the modern metal airplane with his Stout Pullman of 1922, a forerunner of the Ford Tri-motor, designed a rear-engine car, the Scarab, in 1935. Although he only built eight or nine, the car was widely publicized as an ideal streamlined vehicle. Like Buckminster Fuller's Dymaxion Car of 1933, Stout's Scarab more nearly resembled a submarine sandwich on wheels than a beetle. Its interior shape was more like a modern minivan.

Stout knew that aerodynamics were more complex than the advertisements about fish and birds. He spoke instead of "terranautics." Explaining the shape of his Scarab, Stout emphasized the effects not just of the head-on wind resistance that wind tunnels tested but of cross winds—something to which later Bug drivers could testify. But Stout thought that sleeker lines would make his car look modern—and therefore sell better.

Automobile designers began to argue that aerodynamic shapes were natural and therefore beautiful. If it works right, it will look right, was the old creed of the shipbuilder, but in the context of emerging modernist theory it took on new connotations.

Not just the shapes of biology inspired designers during this period.

The Beetle emerged at a time when modernist architects and designers were considering how to employ mass production to bring well-designed, affordable products to the mass market. Inspired by social as well as aesthetic ideals, they rethought designs for furniture, lighting, and household objects as well as housing. That the automobile should come in for the same treatment seemed inevitable. Although Hitler had shut down the Bauhaus, which was the center of modern theory, the Beetle seemed to emerge from the same philosophy of simplicity and reduction to basics. With the passage of time, it seemed to stand alongside such modernist experiments as metal chairs, Bauhaus lamps, and garden apartment complexes.

The advertising for the Beetle would later foster the notion that it was an undesigned object, shaped purely by function. This was the greatest of modernist myths. It was an answer to the questions: what are the shapes that persist, beyond fashion or style, and why do they persist?

Objects were beautiful as they became more functional, modernists argued. Modernism believed in the power and persistence of the Platonic solids—objects without movement, beautiful by virtue of sheer geometry. It was the most fundamental of modernist ideals.

In the introduction to his landmark Machine Art show at the Museum of Modern Art in 1934, the year of the Airflow Chrysler, Philip Johnson quoted from Plato's *Philebus:* "By beauty of shapes I do not mean, as most people would suppose, the beauty of living figures or pictures but . . . straight lines and circles, and shapes, plain or solid, made from them by lathe, ruler and square. These are not like other things beautiful relatively but always and absolutely." As examples, Johnson offered up the functional beauty of gears and propellers and ball bearings. The same beauty might have been found in the Bug's engine, but not in its body, whose aerodynamics were closer to the sleek, streamlined Lincoln Zephyr in which Johnson crisscrossed Europe in the 1930s.

Was the Beetle, in this sense, modern? Modernism urged function as the source of form, but when modernists took to cars it was with rulers, not curves. Le Corbusier's *Voiture Maximum* was like a box, a plywood assemblage—more of a voiture minimum. The Bauhaus's founder, Walter Gropius, hired to design a body for an Adler automobile, came up with a boxy design that ignored every existing lesson of the wind tunnel. The contrast between the flowing shapes of moving objects and the static ones of Bauhaus geometry was profound.

When streamlining was quickly applied to static objects, where it had no function, modernists began to attack it. The best known industrial designer, Raymond Loewy, designed a streamlined pencil sharpener that, although never produced, became a notorious example of nonfunctional aerodynamics. Streamlining, the Museum of Modern Art declared, was not modern design. But the industrial designers who were capturing the fancy of the public and corporate America saw streamlining as inevitable. "Most advanced yet acceptable" was Loewy's slogan. He sketched the cars of the future in ten-year intervals, imagining increasingly streamlined vehicles whose evolution suggested that of an animal species.

Aerodynamics only come into play at speeds above 40 or 50 miles per hour, a speed at which the Volkswagen, with its mere 25 horsepower motor, would spend little time. For all its rounded lines, the Bug with its upright windshield yielded only a .38 coefficient of drag, good for its era, but unimpressive today when even the least-streamlined-looking cars score .25 or .27.

How much sleeker and wind-cheating it could have been was suggested in 1938 when Porsche designed an alternative body to create a single-seat sports car for a planned Berlin-to-Rome motor race. The first 400 miles were to be run on the autobahn between Berlin and Munich, and Porsche calculated that a version of the Bug chassis covered with a body like a Silver Arrow racer, but with only a 50 horsepower engine, would have cruised at 100 miles per hour. Porsche never had a chance to prove the virtues of this Type 64. The race, scheduled for September 1939, was superseded by other events. Porsche drove the car around Stuttgart for a time.

Streamlining and aerodynamics had been a fascination of car designers since the 1890s. The first results were a handful of unique, strange cars such as the Rumpler Teardrop in Germany, the Burney Streamliner in Great Britain, and later the Dubonnet in France. But with the metal body revolution around 1930, when stamped steel replaced a mixture of wood and metal and sometimes even canvas, designers began to give more attention to shaping cars for wind resistance.

Hitler blamed the Treaty of Versailles for all kinds of injustices to the Germans, but he never gave it credit for helping make the country a center for aerodynamics. Forbidden by the treaty to build up its air force, Germany used its wind tunnels to study other vehicles. The more the engineers studied, the further away the results were from the idealized view of man-made moving objects in the shapes of birds and bugs. Because of this research, the Germans literally wrote the book on the subject, called *Aerodynamik des Kraftfahrzeugs,* or *Automobile Aerodynamics.* It was written in 1936 by an aristocratic engineer and enthusiast author named Rheinhard Freiherr von König-Fachsenfeld. The book's ideas would help shape Germany's Silver Arrow racers and speed record cars.

Paul Jaray and Wolfgang Klemperer, aerodynamicists at the Zeppelin works before the war, arrived at what they believed to be the ideal shape for a basic aerodynamic, low-drag automobile, which Jaray patented in most European countries and the U.S. between 1921 and 1927.

Another view came from Stuttgart University, located close to the center of the auto industry. There, in 1935, Dr. Wunibald Kamm found that the key to reducing drag was not tapering the rear end of a car like a teardrop. To deter the formation of air disturbance and drag behind the car, the best shape was a back with a kind of lip on it to detach the air flowing from the body. The "Kamm back"—today the term is generic—was not naturally beautiful but it was efficient.

Many major manufacturers licensed Jaray's patents and studied König-Fachsenfeld's book, but none did more with them than Hans Ledwinka. He was the designer of the Tatra, the Czech car that Hitler's supposed sketches most resembled, and an innovator in automotive design and engineering on a par with Porsche. VW was forced to pay royalties for features Tatra had patented first, and Porsche noted later that "we looked over each other's shoulders."

Tatras, not unlike the Bug, featured spinelike central structures and rear-mounted air-cooled engines. The Tatra 77 model, with a coefficient of drag of .19, could make 93 miles per hour out of just 65 horsepower. Even the underbody was streamlined. The larger Tatra 87, introduced in 1937, was considered the ideal car for the autobahns. Dr. Todt himself owned one. Thanks to a special dispensation, production continued even after Germany invaded Czechoslovakia; it was probably the only automobile that was made without interruption throughout the war.

Ledwinka, like Porsche, believed that putting the engine in back produced more even distribution of weight, which improved handling. But like all rear-engine cars, the Bug included, the Tatra 87 was tail-heavy and given to oversteer, causing its rear end to slide out in turns. The Tatra was a favorite of German military officers, but so many were killed in crashes that it became known as the Czechs' revenge, and SS officers were finally forbidden to use it.

The Tatra was in many ways technically ahead of its streamlined American contemporaries, the Airflow or the Lincoln Zephyr, but by 1940 streamlining was being adopted in the United States in at least token form in a vast variety of cars, vacuum cleaners, refrigerators, irons, and toasters. For Americans, the vision of an aerodynamic future was at once romantic and commercial. It represented a form of modernism more natural and acceptable than the rectilinearity of the Bauhaus.

For American manufacturers, streamlining celebrated their power and efficiency and symbolized the country's economic revival. For Europeans, the wealth of consumer goods, including automobiles, was but a distant dream.

chapter five

I n the 1920s Europeans became fascinated with American industry and engineering and the wonders they produced—huge bridges and skyscrapers, even grain silos and dams. They were eager to visit Ford's Highland Park and River Rouge plants in Michigan.

Germans, in particular, were obsessed with the American industrial boom of the 1920s. They were languishing in a combination of slow economic growth, inflation, and class hostilities. Polarization between right and left exploded in street brawls. Germans seized on the ideas of the American industrial efficiency expert Frederick Taylor as a possible way out and the example of Henry Ford as a source of something like spiritual regeneration.

In his 1927 book *There Stands America,* a German engineer named Otto Moog referred to Ford as "the automobile king." Visiting the Highland Park factory where Model Ts were built, he declared that "no symphony, no Eroica compares in depth, content and power to the music of the factory." Moog declared himself "overwhelmed by this daring expression of the human spirit."

For European architects, engineers, and businessmen like Fiat's Agnelli and Peugeot of France, but especially for the Germans, the spectacle of the factories turning out Ford's people's car came to stand for the new vision of America. It represented the economic miracle, the Wirtschaftswunder, that would describe Germany in the 1950s.

Opened in 1931, Ford's River Rouge factory was even larger than Highland Park and kindled competing visions in the minds of its visitors. Barges and trains brought vast piles of sand and iron ore at one end, and at the other finished cars emerged. The American realist artist Charles Sheeler photographed and painted the factory over and over again, seeing in it the embodiment of his precisionist aesthetic.

The Mexican muralist Diego Rivera was fascinated by the Rouge plant, and in 1932 accepted the patronage of Henry's son, Edsel Ford, to paint murals at the Detroit Institute of Arts celebrating the factory as the spirit of the city and the country. The result combines the folkloric quality of the best of his Mexican work with a sharply observed realism reflecting months of sketching the workers. It also expresses the romance many visitors felt about the place.

One section of the murals shows production of the brand-new V-8, soon to be legendary as outlaw Clyde Barker's favorite power plant. (Clyde and Bonnie wrote Henry Ford a grateful letter.) The orange glow of the foundry radiates light into the whole space while the tilted bucket of a Bessemer steel converter hovers like a primitive idol. A vital tension energizes the angles of leaning and bending workers and seems to animate the machinery itself. The workers are clearly featured individuals and their various ethnic and racial backgrounds are distinguishable in their faces and clothes—even in their hats. Here were socialism and capitalism, labor and management sharing in the romance of mass production as a primal force.

Rivera may have been a friend of Trotsky but he liked Edsel Ford, his capitalist patron, and considered him an artist in his own right. The erstwhile socialist seemed to have worked out, visually and perhaps wishfully, a reconciliation between the forces of labor and capital that in the real world were locked in opposition. Rivera did not gloss over labor's unrest—a stern supervisor on the assembly line, clipboard in hand, looks like a gangster. In fact, River Rouge's assembly line was still tough on workers, Ford's intrusive "sociology," or personnel, department bordered on fascistic, and the company refused to recognize a union.

Porsche's fascination with the U.S. was shared by many Europeans. It was no accident that Kafka's last book, posthumously titled *Amerika,* sent a young man from Central Europe into a strange new world where skyscrapers and giant cactuses seemed to inhabit a single landscape before ending up in the baffling mystic carnival called the "natural theater of Oklahoma." Kafka's geography is jumbled but it doesn't matter; for Europeans, the very American names were magic.

While skeptics saw the inequities of the assembly line—Aldous Huxley depicted a dystopia ruled by Fordism in his 1932 novel *Brave New World*— most of the American public viewed Ford's labor relations as favorably as his universal car. He was the father of the five-dollar day, after all, even though insiders knew that his high wages were not just a matter of benevolence. They were required to retain workers who otherwise would not be able to endure the pace of the line.

As late as 1937, after GM and Chrysler had allowed the United Auto Workers to unionize their plants while Ford still resisted, a poll found that a huge majority of the American public still believed Ford offered better conditions for the worker than any other company.

While the VW3s were running through their tests in October 1936, Ferdinand Porsche and Ghislaine Kaes, his nephew and secretary, boarded the German ocean liner *Bremen* for America. Porsche wanted to study Detroit's assembly lines. He would return the next year to look for machinery and men to outfit the factory and to meet Henry Ford. These practical trips to acquire knowledge, equipment, and people to build the Volkswagen also constituted spiritual pilgrimages. For Porsche, the romance of River Rouge and the myth of Henry Ford were inseparable from the vision of the people's car he planned for Hitler.

In New York, Porsche visited Radio City Music Hall, the Empire State Building, and the new motor highways of Long Island, and watched the Vanderbilt Cup race.

By the 1930s costs had forced William K. Vanderbilt to open his Long Island Motor Parkway as a public toll road. Then, in 1938, in an event that marked a symbolic end of the automobile as a luxury for millionaires, he was forced to donate the road to the state of New York in lieu of $90,000 in back taxes. Robert Moses later incorporated it into his system of public parkways dedicated to leisure motoring for the middle class. Around the same time, work began on the autobahn-inspired Pennsylvania Turnpike, a "defense highway" and the first true American superhighway. The impresarios of the 1939 World's Fair were preparing their visions of the magic motorways of tomorrow, with idealized future cars for all, even if Hitler's autobahn plans were years ahead of the most progressive U.S. highways.

By the time Porsche attended the race, the Vanderbilt Cup had moved to the brand-new Roosevelt Raceway, in Westbury, Long Island, not far from Roosevelt Field, where Lindbergh had departed nine years earlier on his flight over the Atlantic. The Cup organizers wanted European participants to legitimize their event. Auto Union and Mercedes not only received government bonuses for top finishes in races but they knew that a victory in the U.S. would be especially prestigious after Germany's humiliations at the 1936 Olympics. Porsche immediately began thinking about entering cars the following year. On October 12, Columbus Day, Porsche watched as the Italian Tazio Nuvolari easily won the 300-mile race in an Alfa Romeo.

He traveled on to Detroit by train, checked into the Book-Cadillac hotel, a copper-roofed tan pile, and visited the factories of Lincoln, Packard, and General Motors.

Detroit's manufacturers, flattered and curious, rolled out the red carpet not only to their factories but their labs and back studios. At Chrysler, he was shown a small streamlined car called the Star. He took notes and measured and calculated the efficiency of the plants, the number of workers in each, and the man-hours that went into each automobile.

On his return trip east, he stopped in South Bend and Cincinnati and bought a six-cylinder Packard for $1,000. He observed the U.S. elections on November 4, then boarded the *Queen Mary*. He stopped in Great Britain to visit the Austin plant in Longbridge, and was home in Stuttgart by the middle of November, with Kaes driving the new Packard.

Porsche made a second trip to the U.S. in 1937 and began to hire skilled engineers and mechanics for the Volkswagen factory. They were men like Joseph Warner from Ford, who began working for VW on September 1, 1937, and ended up the head of the VW factory after World War II.

At the Vanderbilt Cup race that July, Porsche's Auto Union car, with Berndt Rosemeyer at the wheel, brought home not only $122,100 in first-prize money but a welcome propaganda coup for Hitler. Hitler made Rosemeyer an officer of the SS, an honor the apolitical and skeptical Rosemeyer privately scorned.

Porsche and his party, including Nazi officials, went on to Detroit where Porsche at last met Henry Ford. "What will the car of the future be like?" asked young Ferry Porsche, who had come with his father. "Let him tell me," Ford is supposed to have said, turning to the elder Porsche. "It is for the young to accomplish."*

Porsche invited Ford to Germany, but he politely declined; war was coming, he was sure. Porsche seemed genuinely shocked: how could war be possible in an age of advanced technology? Ford did not answer, the story goes, but moved on to his next audience, with a group of orphans.

There were plenty of other warnings about the future for both Porsche and Ford, if they had chosen to read them. On May 6, around the time Porsche arrived in the U.S., the German dirigible *Hindenburg* crashed in New Jersey. The airship had been filled with hydrogen because the U.S. had refused to sell the increasingly menacing German government helium at prices it considered reasonable. Ford, already under fire for his support of anti-Semitic publications, began to lose public favor for his anti-union stance. On May 27, the myth of labor's satisfaction under Ford was shattered by the "Battle of the Overpass," in which thugs employed by Harry Bennett, Ford's personnel chief, beat up leaders of the United Auto Workers on a bridge just a few feet from the River Rouge plant. General Motors recognized the union that fall; Chrysler soon followed. Ford hung on until 1941 without coming to an accommodation with the union; his reputation as the friend of the worker was gone.

* This tale has the ring of Ford's PR department based on his many audiences.

In addition to visiting factories and shops, Porsche's contingent bought American machinery. Noting that the latest American car trend was that the doors were universally hinged at the front, not at the back as on the Bug prototypes, he ordered the Volkswagen to follow the new style.

Porsche calculated how to produce the Volkswagen that would meet Hitler's price limit. The thousand-mark maximum price tag translated to roughly $500 in the 1930s, a price lower than that of the Model A or any other American car of the time. Today, efficiency in manufacturing is figured as man-hours per car or cars per worker per year. Then the common standard for calculation was cost per pound of production. Porsche concluded that the Volkswagen workers—12,000 in two shifts—could turn out Hitler's little car for less than 1,000 marks.

General Motors's factories were not yet as efficient as Ford's one-model factory at River Rouge, but they had added something to the equation: variation. They were producing Cadillacs, Chevrolets, Buicks, and Oldsmobiles by taking the best of Ford's ideas and building on them. The single model, the people's car, was vanishing.

The end had come suddenly for the Model T. By 1927 it was nearly twenty years old; sales had dropped dramatically to only a third of the year before. A year later, Ford introduced the Model A. But Ford's universal car, the car that rarely changed, the car for the multitude, would last only a few years. When Porsche and his party bought a Model A to try out for themselves in 1937, it was already obsolete.

General Motors chief Alfred Sloan was the real prophet of the auto industry in the 1930s. By 1937 Sloan's "mass class" approach was supplanting Ford's mass production one, just as the GM ideal of "a car for every purse and purpose" was replacing Ford's "universal car" model. Nothing better symbolized the change than the brightly colored cars rolling out of General Motors plants. (One of GM's major innovations was the development of rapid-drying Duco lacquer.) No longer was the buyer stuck with black.

The Model T was a relic and the Model A represented the last gasp of the universal car in America. Soon the ideal of River Rouge and the assembly line was under attack, too.

Alfred Sloan saw a car not just as a useful transportation tool for farmer and worker, but as a marker of status and personal expression. Ironically, once owning a car had become a necessity, offering a choice of models became a necessary luxury. An automobile company, Sloan reasoned, should offer variety—"a car for every purse and purpose" in his much quoted phrase. Changing the array of models and offering them in a range of price, features, and style would create discontent with last year's cars.

Ford had long ago summarily rejected the idea of what was to be called planned obsolescence. In *My Life and Work,* he had declared that "it is considered good manufacturing practice, and not bad ethics occasionally to change designs of the old models." But "we cannot conceive how to serve the customer unless we make for him something that as far as we can provide will last forever. . . . It does not please us for buyer's hardware ever to become obsolete . . . we want the man who buys one of our products never to have to buy another."

It was a noble sentiment, but one the market had outgrown.

GM had taken on Ford with its Chevrolet 490, named for its price and marketed on the claim that it offered more than the Model T. But for Sloan the real people's car was a used car. And used cars were created by pushing new cars.

Continued sales of new cars depended on financing purchases, or buying on time, another sales tool Ford had emphasized. Credit was a natural mode of doing business in optimistic, expanding America, and had been used to sell sewing machines and other items since the 1880s. It caught on in earnest in the automobile market in the 1920s when the prospects for expanded incomes were rising like the stock market. As much as the Model T, buying on time brought cars to the American people.

Combined with finance purchasing, the annual model change was vital to Sloan's plan. This year's new car became next year's used one.

Thus the auto business began manufacturing discontent, not just cars. Design and styling were vital not only to distinguish the top from the bottom of the line (though the chassis and engine might not vary so much), but to distinguish next year's model from last year's. Upward mobility was linked to forward mobility on what Daniel Boorstin called "the ladder of consumption." As early as 1926, Sloan was concerned about the effect of appearance and design on sales. The Model T was tall and gawky. Sloan wanted to lower cars to make them look fresh and modern. He worried about "beauty of design, harmony of lines, attractiveness of color schemes and general contour of the whole piece of apparatus." A cold and cadaverous figure, he seemed to be the last man who would understand the importance of style and the least likely to hire "Hollywood" Harley Earl, the father of American automobile design, to do something about it. Fittingly, the GM Art and Color Section that Earl headed was established in 1927, the last year of the all-black Model T.

In Hollywood, Harley Earl had worked with the custom coach builder Don Lee pioneering the use of modeling clay to create the forms. He built cars for the stars and lived next door to Cecil B. DeMille, a figure with whom he shared striking similarities.

For GM Earl created a system of studios, just as the film business had its studio system, and over the years he got great work from designers while both abusing and channeling their individual talents and impulses. Like the Hollywood studios, Earl's created some of the greatest and most characteristic artifacts of American culture. To him, cars were diversions, entertainment—like movies.

"You can design a car," he proclaimed, "so that every time you get in it, you have a little vacation for a while."

In a famous *Saturday Evening Post* article of 1954, Earl declared that "my primary purpose for 28 years has been to lengthen and lower the American automobile, at times in reality and always at least in appearance. Why? Because my sense of proportion tells me an oblong is more attractive than a square . . . and a ranch house more attractive than a 3-story house and a greyhound more attractive than a bulldog." General Motors automobiles in 1926 stood seventy to seventy-five inches high; by 1963, when Earl was finished, they stood only fifty-five inches above the road.

Earl's philosophy reinforced the role of the automobile as a marker of social status. In contrast to the generic people's car, his were personal cars.

Porsche's visit to Ford strengthened public suspicions that the American automaker supported the Nazis. In 1937, responding to public outcry, Ford had disavowed support for the anti-Semitic publication *International Jew*. But his link to Hitler led to stories that Ford had financed the Nazis during their rise to power.

The Nazis had asked Ford to open a plant in Berlin to build three-ton trucks with clear military utility. Despite initial reluctance, in April 1938, Ford agreed but only to make ordinary trucks and automobiles. On his seventy-fifth birthday, July 30, he accepted the Grand Cross of the German Eagle in his Dearborn office. Hitler bestowed the medal ostensibly for Ford's role in "giving automobiles to the masses" but implicitly it was to reward him for opening the factory. By the time it went into operation in 1939, war had put it beyond Ford's control and the Nazis were free to build military trucks.

In 1938 VW representatives in America to recruit more men and machines for the plant found they were no longer welcome. That November crowds gathered outside the German consulates protesting the events of Kristallnacht.

Porsche did not have to go all the way to the U.S. to see a Ford plant. He could and probably did visit the factory Ford had opened in Cologne in 1931. But there he would have found a very different system than at River Rouge, one that would have served as a warning of how formidable it would be to create a German equivalent of River Rouge.

Ford had tried to transplant his system to Europe as early as 1914, when he opened a plant in Dagenham, England. But he did not succeed in replicating River Rouge there or in plants in Paissy, France, or Cologne.

Efforts to institute assembly line methods in Europe ran afoul of the traditional guild- and craft-based system, which used a piecework rather than hourly pay scale. In the U.K., the Morris and Austin factories encountered similar difficulties. Managers could only deal with assembly line workers through shop stewards. Workers resisted the fragmentation of their jobs that the assembly line required. Sometimes, even when not unionized, they struck over the issue. Only in Fiat's vast Mirafiori plant, constructed in 1936 to build the Topolino, the Italian people's car, was something like the American system achieved.

There were further obstacles to the assembly line. Europe lacked the infrastructure of parts and materials suppliers found in the United States. Neither did it share the American tradition of interchangeable parts, which extended back to early-nineteenth-century gunmakers. Interchangeable parts had been used by U.S. luxury automakers like Cadillac well before Henry Ford combined the idea with the moving assembly line. When General Motors bought Opel in 1925, Alfred Sloan recalled, German dealers resisted the system because they made so much money fabricating replacement parts on the spot. It was not clear that the assembly line model would travel across the Atlantic at all. Building the Volkswagen would mean not just building a new car or a new factory but a new industrial culture.

b y May 1937 when Porsche left on his second trip to the United States, impatience for real evidence of the Volkswagen was mounting in Germany. One joke compared the Volkswagen to Christ—everyone talks about it but no one has seen it.

Hitler was losing patience, too. To open the Berlin auto show in February 1937, Germany's leading race car drivers assembled in the courtyard of the Chancellery and drove to the show hall through streets lined with 12,000 storm troopers. At the show Hitler banged his fist and declared, "Either German automobile makers produce the cheap car or they go out of business. I will not tolerate the plea it can't be done." He expressed disappointment at such efforts as Opel's P4. At 1,430 marks it was an inexpensive model, but much more than the price the Führer demanded for the people's car, and he dressed down young Opel manager Heinrich (Heinz) Nordhoff.

The automakers were on the spot and the RDA, their trade organization, sought a way out. There was no question, given the political pressure, that the RDA's evaluation of the VW tests would be favorable, but the team in charge of the evaluation, led by Wilhelm Vorwig, was genuinely impressed with the car. What was unclear was who should build it—and how the RDA could protect its members from the potential competition it represented. Porsche himself, in his proposal to Hitler in 1933, had left open the question of who would build the car he proposed to design and test.

Faced with the inevitability that the Volkswagen would be built, the bosses of the RDA, director Wilhelm Vorwig, Count von Opel, and BMW chairman Franz Josef Popp, sought the most favorable plan for the member manufacturers. The RDA offered to set up a company to manufacture the car—if provided with a huge government subsidy. But Hitler had tired of the automakers' reluctance to fall into line and no longer trusted them. For their part, the RDA's directors reasoned that a new, government-established company might draw resources from the private sector and if any single manufacturer took on the Volkswagen project with Nazi support, it could threaten the other automakers. The RDA chiefs hit upon the idea of handing the project to the German Labor Front—Deutsche Arbeitsfront, or DAF—the state union created by the Third Reich to replace the private labor unions it ruthlessly suppressed. After celebrating the first of May as "the day of national labor" in an effort to co-opt the traditional holiday of the left, the Nazis outlawed the trade unions on May 2, 1933, seized their offices and treasuries, and jailed even pro-Nazi union officials. The DAF proclaimed the end of collective bargaining and imposed a ban on strikes. The regime promised to make the employers once more "master of the house"—a phrase of the time that appealed to big business.

The DAF posed the least threat to the private makers. The RDA argued that if, as the Volkswagen's proponents said, the buyers for their car would be working people who could not afford existing models, then why not let the organization dedicated to workers sell the car through its network of local branches? The DAF had already produced such government-mandated products as the Volksempfänger, the inexpensive radios the regime promoted to extend the reach of Hitler's and Goebbels's speeches. Let the creators of the people's radio produce the people's car.

The RDA also feared that the Volkswagen would compete with its members for scarce capital. But the DAF, enriched by dues from workers, had amassed a vast treasury from which it could finance Volkswagen production without drawing on private capital. The RDA would be able to wash its hands of responsibility for it and concentrate on building traditional cars—a business that had at last begun to revive in Germany.

In January 1937, the RDA raised the idea of handing the VW project over to the DAF with Joseph Goebbels, the propaganda chief, who liked it. He knew Hitler would be pleased. Dr. Robert Ley, head of the DAF, was also intrigued but wary as well: there were dangers in taking on a project that was so important to the Führer.

Although less well known than Hermann Göring or Goebbels, Robert Ley was one of the most powerful men in the Reich. Steelmaker Fritz Thyssen called him "a stammering drunkard and womanizer," but he was also a skilled survivor of Nazi infighting. Had he been a more sober man, he would never have dared to take on the task of building Hitler's car.

In 1938 Stephen Roberts reported that "it is extraordinary how powerful the doctor is in Germany and how little known he is outside the country. . . . This man is one of Hitler's oldest personal friends and enjoys license that makes him stand out even in the land of dictators, although his behavior is often a sore trial to the more puritanical Führer himself." But Hitler showed a surprising tolerance for Ley's behavior.

Ley joined the party early, in 1923. Like his leader he had fought in World War I, spent three years in a French POW camp, and carried the scars of the experience with him the rest of his days. Returning to civilian life, he sympathized with Hitler's claim that the German army had been stabbed in the back. He worked as a chemist at Bayer, but lost the job due to drunkenness and anti-Semitic activity. In the infighting among Nazis in the 1920s he defended Hitler, the former corporal, with fervor. Partly in gratitude for his establishment of the anti-Semitic publication *Westdeutscher Beobachter,* Hitler put him in charge of the Cologne party and, in 1933, made him head of the DAF.

In that job, Ley wrapped his suppression of the unions in the trappings of labor solidarity. When the DAF took over, Ley intoned, "It was just as if the leaders of the trade unions had waited for them to be taken over, and breathed a sigh of relief when finally relieved of their burden." He unctuously declared, "Workers, your institutions are sacred to us National Socialists. I myself am a poor peasant's son and understand poverty. . . . I know the exploitation of anonymous capitalism. Workers! I swear to you, we will not only keep everything that exists, we will build up the protection and rights of the workers still further."

Workers were forbidden to change jobs without permission. Real wages shrank and heavy assessments for various political contributions cut into their earnings. Membership was required of all workers—and all employers. In the ideal new Germany, there was no distinction between worker and manager. The pre-Nazi unions had counted five million members; the Labor Front now boasted 26 million of the total German population of 66 million. Since all members were required to pay significant dues, the DAF became an important source of cash even the Führer had to respect. The DAF had collected some $200 million by the time the war began.

The DAF established an office to boost workers' morale through a combination of propaganda and diversion, called *Kraft durch Freude,* Strength through Joy. The KdF ideal was summed up in the propaganda pitch that held that "the contentment of the day's work vibrates into the leisure hours in which fresh strength is gained for the next working day."

The KdF sponsored sports, theater, and music on a huge scale, paid for by worker contributions. It built a huge resort in the coastal town of Rügen, six cruise ships, and eighty recreation centers across Germany. KdF sports programs involved seven million people; its symphony orchestra traveled around the country. The organization offered $25 cruises and winter skiing excursions to the Bavarian Alps for $11. Workers put aside a few marks each week from their paychecks. "[KdF] buses were to be seen in all corners of the country," Roberts noted, "blocking traffic while disgorging happy worker tourists."

In wall-sized color charts, KdF officials minutely tracked the participation of workers in every region of the country, in activities from theater to music, lectures to hiking, skiing to sailing.

Non-Aryans, of course, were ineligible.

The KdF's greatest triumph was the people's radio, the Volksempfänger, which Hitler had called for along with the Volkswagen in his January 1933 speech.

At the radio trade show in Berlin that August, Goebbels was able to tout the industry's cooperation in producing the first model, a brown box rounded at the top with a circular speaker designed by Walter Marie Kersting and produced by several firms in Kiel. Its model number was VE 301, after the European form of January 30, 1933, the date Hitler took office. A widely reproduced poster pictured the radio as a gigantic brown gothic temple, towering above a crowd. "All of Germany listens to the Leader with the People's Radio," was the ad slogan.

A million and a half sets at 35 marks each were sold the first year. Factories and offices were also equipped with sets. For households without one, the Nazis filled the streets with loudspeakers. By May of 1939 some 70 percent of the population was able to hear Goebbels's harangues. Radio was "bringing German culture to the entire German people regardless of their income," he claimed. "Thanks to the people's radio set, a solid inexpensive and capable receiver, the number of radio listeners has risen from around one million in 1932 to 9.1 million today. The un-German programming of the system era"—a euphemism for the Weimar Republic—"has been transformed by National Socialism. Now radio acquaints the German people with the work of their great masters of musical literature. Alongside these artistic programs, entertainment programs provide for the relaxation of hard-working people."

Ley and his underlings dreamed of turning out people's versions of other products, from housing to appliances, refrigerators, and vacuum cleaners. A people's refrigerator was never realized, but in the United States, General Motors came close to providing it with the Frigidaire.

GM executives treated refrigerators like cars. Both were "metal boxes with motors," and like the first widely affordable cars, the first refrigerators were very basic. But diversity soon followed: To streamline its Coldspot model, Sears called in industrial designer Raymond Loewy, who rounded its shape and added chrome strips. Soon whole product lines were developed, arranged by features, style, and price, as in the Sloanist automotive market.

The people's radio monopolized the German market, and Nazi propaganda often cited it as one of Hitler's achievements. The car manufacturers had not responded to his demand for a Volkswagen, and now those who had made the people's radio would make the people's car.

In May 1937, the government formally established a company to build the Volkswagen called the Gesellschaft zur Vorbereitung des Deutschen Volkswagens. Its acronym, GeZuVor, meant "go forward." GeZuVor would be financed by the DAF. Hitler planned to build a million and a half cars a year.

The DAF's first job was to find a location for the factory to produce the car, and Ley's aide at the DAF, Bodo Lafferentz, prospected for a site from a Ju-88 bomber. The complex would be as big as Henry Ford's River Rouge and would eventually include its own steel and glass mills and a plant to make Buna, or synthetic rubber.

In the Lüneburg Heath area of Lower Saxony, Lafferentz found a spot that offered proximity to rail, autobahn, and canal. The flat, sometimes swampy, area was lightly inhabited. A few scattered towns, with houses grouped around castles and churches, dated to feudal times. Most of the area was held by large estates that remained in the hands of a few owners, many with aristocratic titles that went back centuries. It had been the site of battles during the religious wars of the Reformation, and in its forests and swamps, strange stones had been found showing evidence of primitive sacrifices of animals.

Official cars began appearing in the area in the fall of 1937. Locals thought they must be there surveying for the new autobahns. The planners focused on a huge tract of land that had been in the family of Count von

Schulenberg since it was granted to his ancestor by Emperor Lothair the Second in 1135 as a reward for defending the area against Slavic invaders.

The Count, however, resisted. He tried to pull strings in Berlin, but succeeded only in winning a delay and at last was forced to sell. And in a weird incidence of Third Reich environmental sensitivity, a government ministry found that some seventy species of gnats were endangered by plans to build on the heath.

In early 1937, in what had by now become an annual ritual, Hitler presented more news of the Volkswagen at the Berlin auto show and unveiled a vast white plaster model of the factory.

Fallersleben, the largest town near the site of the Volkswagen factory, was a cozy, gemütlich town, a fairy-tale set of twisting streets, built in the sixteenth century by Duke Franz von Brunswick und Lüneburg. On May 8, 1938, breezes gently billowed the curtains in the open windows of the half-timbered house where the popular poet Heinrich Hoffmann von Fallersleben, author of the national anthem, had been born in 1798. The house was decorated with wreaths of laurel and small flags bearing the swastika.

Fallersleben and its inn accommodated the dignitaries who arrived on May 26 for the cornerstone-laying ceremony a few miles away, on the other side of the Mittelland Canal.

Dr. Robert Ley, in his role as head of the German Labor Front and its subsidiary, the Strength through Joy organization, had issued invitations weeks before and provided maps showing the routes of the express buses and the twenty-eight special trains. The new autobahn was illustrated on the invitation, complete with its curving exit ramps and parking areas.

A tall brick reviewing stand had been constructed at the site and hung with swastika banners interspersed with fir boughs. Workers had been given the day off, bands entertained the assembled contingents of Hitler youth, SA, SS, and representatives of the local gauleiter. Himmler, Huhnlein, the head of the NSKK, the national race car driver's union, and Count von Schulenberg were among those standing for hours in the watery sunlight waiting for the Führer, who came by his special train from Munich.

The bricklayers were dressed in old-fashioned suede suits with black top hats, the carpenters in black velvet suits. A shirtless well-muscled young man, the very icon of Aryan labor, straight off a propaganda poster, prepared the stone. Nearby stood Ferdinand Porsche looking distracted and out of place in his rumpled trench coat.

When the Führer arrived some thought he looked drawn and tired. In March, he had ordered the invasion of Austria and the "Anschluss" creating "Greater Germany." For weeks he had made shrill demands for the rights of ethnic Germans in the Czechoslovakia province of Sudeten and two days after visiting Fallersleben, he solidified his plan to dismember the country.

From the station, decked out in Nazi banners, Hitler was driven in his Mercedes limousine to a hill overlooking the site of the future factory. The plan was of a piece with Hitler's other grandiose city building plans. A model town would be tied to the factory and the former castle of Count von Schulenberg along an axis that crossed the Mittelland Canal. The boulevard would be wide enough for martial parades but the neighborhoods were laid out around small squares to foster a sense of community. Ponds, parks, and playgrounds were to be created and forest areas preserved. Hitler wanted the factory and the city beside it—like the villages he planned in the east and the new Berlin—to showcase the ideals of his regime. It would be a model city not just for workers but for the Nazi order.

Hitler the amateur architect was in his element, and he and Albert Speer, the Führer's favorite architect, chose several architectural firms. For the factory, Mewes, from Cologne; Kohnbecker, from Gaggenau; and Schupp and Kremmer, factory design specialists with offices in Essen and Berlin. The city design fell to Peter Koller, a young protégé of Speer's.

The few sections of what would be called KdF-Stadt completed before the war included some 2,400 units of housing, or about a tenth of the planned work. One area, the Steimkerberg neighborhood, housing officials, was nearly finished. It featured three- and four-story houses that could have come from a comfortable bourgeois vision of early nineteenth-century Germany.

With Germany rearming and building going on elsewhere, there weren't enough workers for the project so Mussolini sent thousands of unemployed Italian construction workers. The factory rose quickly but most of the site remained a lake of mud, supporting an archipelago of barracks and temporary buildings clustered around the great ship of the factory.

After all his frustrated efforts to build a people's car, Ferdinand Porsche seemed at last to have found a client who wouldn't fail him.

Porsche demonstrated his commitment to the Volkswagen by moving from Stuttgart to Fallersleben even before the factory was finished. He built himself a wooden bungalow in the forest near the factory and surrounded it with hundreds of tulips. When not working, he hunted on the nearby heaths. His son-in-law and lawyer, Anton Piëch, soon became head of the factory and his son, Ferry, continued as his chief aide.

Porsche commonly addressed the dictator simply as "Mr. Hitler," and would stiffly and routinely extend his arm in the Nazi salute. Ideology bored the engineer as much as ceremony, but he and his staff knew how to ingratiate themselves.

The month before the cornerstone ceremony, on Hitler's birthday, Porsche's technicians presented the Führer with an amazingly exact one-tenth scale model of the Beetle complete with rubber tires. Photos show a delighted Hitler examining the model as Göring and others look on.*

The Volkswagen project increasingly kept Porsche away from working on the racing cars, but the Silver Arrows were employed more and more in propaganda. Hitler pushed Auto Union and Mercedes to build ever more streamlined models so he could claim the world land speed record for Germany.

While other speed record attempts, like those of Sir Malcolm Campbell and his British Bluebird, were made on the flat, dry lake bed at Bonneville, Utah, Hitler insisted that the German attempts be made on his autobahns. A special stretch of highway between Frankfurt and Darm-

---

* The model has survived and is in the Deutsche Museum in Munich.

stadt was widened to sixteen feet and the median removed but it was still
very narrow for such speeds and the crosswinds and overpasses made it
even more dangerous. The chief Auto Union driver, Berndt Rosemeyer,
said that when he approached the overpasses at nearly 250 miles an hour
they looked no bigger than keyholes. But the lives of the drivers were sec-
ondary considerations. In January 1938, winds swirling around the overpasses caught Rosemeyer's Silver Arrow and virtually peeled the aerodynamic body off the chassis. He was flung from the car and landed in a
nearby grove of trees, sitting upright so peacefully that the men who
reached him thought for a few seconds he was still alive. Hitler ordered
that Rosemeyer be buried with full military honors, but his bitter widow
refused. Porsche was not there; he would have forbidden Rosemeyer to
drive under such windy conditions, he said. He felt the loss deeply, as a
waste, but he remained silent.

At last Hitler arrived at the ceremonial area.

He strode across the open field, his jodhpurs blossoming over the
tops of his high boots, accompanied by an officer with a sword, past the
ranks of storm troopers with their coal scuttle helmets.

For the millions listening on their people's radios came the sounds
of music and applause and the voice of an excited narrator describing the
scene: "He is here! The Führer's official standard is unfurled. He marches
forward, gathers up the flowers from a young girl. He greets the soldiers
reviewing the honor guard with brief salutes."

The speeches got under way. Dr. Ley and his deputy, Bodo Lafferentz,
spoke first.

"It is your work, my Führer," Ley proclaimed. "This great factory and
the city around it are one of the greatest social works of all time and all
countries." They would become "the materialization in stone and iron of
the ideas of the regime. . . . In ten years time there will be no working
person who does not own a people's car.

"What has been started here—this factory and everything which will
come of it," he went on, "is fundamentally and uniquely your work. We

know how you thought of giving the German people a good but inexpensive motor vehicle even before you came to power and how you have ever since imbued with new strength all the designers and others who labored on this car."

Ley pointed across the canal. "Over there we shall build our beautiful city . . . the lovely wooded green landscape offers an opportunity for a superb urban development. The city which will be built here will rank among the most beautiful in the entire world."

Lafferentz and Ley presented the people's car as the Führer's gift to the people. It was not something they deserved or had earned but something he bestowed on them.

They repeated a story invented by Jacob Werlin that had become part of Hitler's legend. Werlin had often told audiences of driving with Hitler one cold rainy night before he took power. Hitler stopped the car to give a ride to a motorcyclist. The man was cold and pitiful, and Hitler, warm and dry in his Mercedes, had said, "If I were in a position to do so I would like more than anything to give every one of these people a car as a present."

Hitler then spoke, striking his familiar theme of motorization of the masses. "When I came to power in 1933," he declared, "I saw one problem that had to be tackled at once—the problem of motorization. In this sphere Germany was behind everyone else. The output of private cars in Germany had reached the laughable figure of 46,000 a year. And the first step toward putting an end to this was to do away with the idea that the motor car is an article of luxury."

Producing a people's car, Hitler continued, would also help make the country self-sufficient in food production. He had often expressed this piece of crackpot economics. The German masses spent too much on imported food, he believed; buying cars would divert those marks to support domestic industry. And, he said, it was "for the broad masses that this car had been built. It is to meet the transportation requirements of these masses and to bring them joy.

"Therefore it shall carry the name of the organization which works hardest to provide the broadest masses of our people with joy and therefore with strength. It shall be called the KdF-Wagen. I undertake the laying of the cornerstone in the name of the German people. This factory shall arise out of the strength of the entire German people and it shall serve the happiness of the German people."

Movie cameras ground away, magnesium bulbs flashed on the big still cameras. Folding his arms over his chest, Hitler watched as two chunky men wearing white suits that looked like pajamas and black top hats—formal wear of the mason's guild—spread mortar over the stone with trowels. The cornerstone itself was a waist-high cube inscribed with the cogwheel and swastika of the DAF.*

Then the cloth covers were pulled off the three sample models of the Strength through Joy cars that were arrayed in front of the stage: the sedan or limousine, the model with the sunroof, and the convertible. The Führer climbed into the front seat of the convertible, with Dr. Ley and Dr. Porsche in the back seat. Ferry Porsche slid behind the wheel and drove, passing the ranks of soldiers and the cheering crowds, to return to the station and Hitler's waiting train.†

When Porsche was told of the car's new name, KdF-Wagen, both he and Ferry were surprised and shocked. The implied ideology would deter buyers at home and appall them abroad, make the car a disaster for potential exports, which Porsche had already been contemplating. But Hitler commanded and the name was made official, and trademarked, as were Volksauto and Volkswagen. Most of the public and world press went on referring to the car as the Volkswagen and it was even stamped as such at the plant. Even SS command cars left the factory with the VW logo, familiar today, and not KdF on their dashboards.

---

* Years later, the stone would be pulled from the site and broken up into concrete for a new bridge over the canal.

† Ferry Porsche had driven the car from the Porsche headquarters in Stuttgart, and was one of the few who knew that it was far from being the sample model convertible. Its engine was supercharged to get 50 horsepower, double the output of the normal model. As usual, the Porsche engineers had not been able to resist making the engine peppier and the car sportier and Ferry Porsche enjoyed the extra power.

The name Volkswagen had been around for a while. The Standard company, a small automaker, had used the name "German Volkswagen" for a model in 1932. In Germany *Volk* had a long history and a special set of meanings. Like the English "folk" applied to folk music or folk art, Volk was wrapped up with romantic ideals of rural people and simple arts, farm life, folk costume, and half-timbered buildings.

But Hitler charged the name with Nazi ideology. *Volk* was critical to Nazi propaganda. *"Ein Volk, ein Reich, ein Führer"*—"one people, one empire, one leader" was a key Nazi motto. The word carried overtones of *"Volk und Boden"*—"people and soil"—and evoked the adjective *völkisch,* or "of the people." It dovetailed neatly with the populist pitch used in KdF propaganda, which flattered its listeners by emphasizing how hardworking they were and subtly ministering to resentment against the upper classes. "The purpose of the KdF organization is to show the less fortunate the wonders and beauties of their nation," Ley said. "It opens opportunities to the people that formerly were reserved for the well-to-do."

Soon, however, "of the people" meant "of the regime." And *volkfremdisch,* "anti-folkish," or against the people, became a propaganda term like "un-American" in the 1950s.

The logo for the car with its familiar V above W was, aptly enough, the work of the man who designed its motor—Franz Reimspiess.

Picking up the hundred marks promised in an office contest, the story goes, Reimspiess drew several variants of the design. All of them fit neatly into a circle, and all echoed the swastika. In one version, the letters lay inside the cogwheel, the gear shape symbol of the DAF. Another version was composed of four arms with winglike shapes, shown here:

The wings are a variant on the swastika, but also evoke the abbreviated brush shapes in electrical diagram symbols for a motor or generator. The logo occupies a special place in design history. Its shape looks back to the graphics of the Vienna Secession, where the flowery letters and images of Art Nouveau were turned into metal and machine shapes. It also recalls the logos of important craft and art groups—the Wiener Werkstätte, the German Werkbund, and the motor company called the Wanderer Werke, for which Porsche had designed a car in 1931—all of which were part of a tradition of interlocking and juxtaposing letters.

The interlocking V and W looked forward to simple model symbols, such as international road signs and Olympic glyphs and icons, as functional and international as the Bug itself was to become. In the 1920s, designers and theorists were already proposing an international pictorial language to cross verbal barriers, a sort of Esperanto of symbols. Otto Neurath developed a well-publicized system he called "isotypes." Eventually, the VW logo would become as recognizable as any of the road signs, Olympic glyphs and icons, or other common symbols Neurath's system inspired.

Most importantly, the logo echoed the rectilinear energy of the swastika. The Nazis controlled the design and use of such graphic symbols rigorously, through "laws for protection of national symbols." This system of symbols, writes Steven Heller in *The Swastika: Symbol Beyond Redemption?*, demonstrated the Nazis' "complete mastery of the design and propaganda processes. . . . Even the most vociferous opponents of Nazism agree that Hitler's identity system is the most ingeniously consistent graphic program ever devised."

The swastika had been put forward as a Nazi symbol in the early days of the party but it was Hitler, the frustrated artist and architect, who refined the colors, tilted it in the circle, and gave it final shape. In *Mein Kampf,* Hitler wrote a brief color analysis of the red, black, and white of the Nazi symbol, calling it "the most resplendent harmony that exists," symbolizing "the fight for the victory of the idea, of creative work which in itself is and always will be anti-Semitic."

Around the swastika grew a system of visual symbols that mapped the mad terror of the whole regime, including such mundane shapes as the cogwheel logo of the DAF and such sinister ones as the runic double lightning bolt esses of the SS. It had its most chilling manifestation in the codes of colors and shapes found in the system of colored triangle badges identifying concentration and labor camp prisoners as Slavs, Poles, leftists, common criminals, and homosexuals, culminating in the yellow double triangles laid over each other to form a Star of David, the traditional symbol of Judaism. These symbols and their placement on garments were specified by the regime with the precision of engineers blueprinting an engine.

The details of the plan for workers to buy the car emerged in the weeks that followed Hitler's laying of the factory cornerstone. The same savings plan used by the KdF, by which workers put aside a few marks a month, was modified for the car. The worker could contribute five marks a week or ten or fifteen and after he had paid in 750 reichsmarks, he would receive an order number for a car. The car could be picked up at the factory. This plan stood in contrast to the American system, in which buying on time was as critical to expanding auto sales as the assembly line.

Some 336,000 people signed on to buy Beetles through the KdF savings plan—far from the million or more cars Hitler wanted to produce each year. But no one ever received a car through the plan. At the end of World War II the Soviets confiscated the 1.5 million marks in the KdF funds, and workers soon filed suit. William Shirer of the International Press Agency, the future author of *Berlin Diary* and *The Rise and Fall of the Third Reich,* called the savings plan "a swindle perpetrated on the German worker."*

* Despite such comments, the failure to deliver on the KdF scheme was not the result of outright fraud. Many business historians have tried to show that the Volkswagen enterprise could never have been profitable. But that is beside the point: had the Nazis avoided war or won the war the regime would likely have produced cars at a loss for the sake of propaganda. Other models might have been added: in his alternative history novel *Fatherland,* which imagines a victorious Germany in 1965, Robert Harris describes four-door VW taxis.

The legal issues were complex: was the West German government liable, or VW itself? What was the relationship between Hitler's Volkswagen company and the postwar Volkswagen company? The matter was not settled until 1961 in an agreement by which savers who had put money into the fund before the war could get either 600 marks in credit toward a new car or 100 in cash. Most opted for the car credit.

The *Times* of London reported the cornerstone ceremony close to a story on a series of air shows taking place across England the following week. *The New York Times* didn't get its story in until July 3, under the headline "German Car for Masses: First of $400 Strength Through Joy Autos Is Expected in 1940." The story referred to the planned car as "the baby Hitler." A photo caption mistakenly credited the car with a plastic body, and the reporter wondered how well the new car would do in a collision. The story compared the cars to "shiny black beetles."

In an atmosphere swirling with propaganda, injunctions to do this or that for the fatherland, the appeals of the KdF-Wagen had trouble standing out. For a time, one local observer noted, there was a "fever" for the Volkswagen, but interest waned and subscriptions to the KdF savings scheme were slow. This was despite such catchy poems as:

> *Fünf Mark die Woche mußt Du sparen—*
> *Willst Du im eignen Wagen fahren!*

Roughly translated:

> *Five marks a week must you put aside—*
> *If in your own car you want to ride!*

Public skepticism about all government schemes demanded special efforts from the promoters of the KdF-Wagen. Children were given a brightly painted metal savings bank to accumulate pfennigs toward the family car. A commemorative postage stamp honoring the KdF-Wagen was issued. There was even a family board game, where players tried to advance their KdF-Wagen rapidly through autobahn stretches or lost time and turns with flat tires. "Educational . . . sociological" proclaimed the game's box.

The promotional literature explicitly linked the car to the Silver Arrow's "racing heritage" of independent suspension, torsion bar, and the unconventionally placed air-cooled engine. And Ferdinand Porsche, whose celebrity was a selling point, stood behind both cars.

Many promotional items featured idyllic visions of families cruising down the autobahns and magical visions of Bugs turned translucent, exposing their mechanical elements and passengers and seeming to hover in the air.

Echoing both Hitler and Henry Ford, who had presented the automobile as a means for the worker to take his family on weekend excursions to the country, much KdF publicity featured picnicking families.

The car's promoters also produced a very clever mechanical manual, organized by a thumb key, with waxy overlays, that highlighted vital parts of the automobile. One view shows a cutaway of the car with five happy passengers. Tucked behind the rear seat and in the front is an impressive array of baggage: a briefcase, a woman's handbag, a hiker's backpack, even a round candy box—a gift to be bestowed at the end of the trip?

A mustached man in coat and tie is at the wheel, eyes fixed on the horizon; a woman in the passenger seat dozes. But in the back seat an older woman (Granny?) and two teenagers stare eerily back at the viewer, the boy's arm around the girl.

Another widely reproduced promotional image depicted a sleek, black KdF-Wagen, speed streaks streaming off its bumpers and hubcaps as it raced across a new autobahn whose concrete is bright and clean. Sunlight brightens the grassy shoulders and a line of snow-topped mountains is visible in the distance, along with three trees—a landscape that could have been quoted from the romantic spiritualist painter Caspar David Friedrich. A boy looks out the rear window, his cherubic face pressed to the glass and his hand raised in greeting.

In jarring contrast to the modern machine, the lettering is Fraktur—the old-style Gothic-based type the Nazis brought back to combat modernist sans serifs—a *Fachwerk* of typography.

Representatives of the international press were invited to test-drive the Volkswagen in the spring of 1939. They were generally impressed. The reporter for Britain's *Motor* magazine noted that the car was noisy but praised its ability to maintain a steady 50 miles per hour even uphill. But these cars were hand-built prototypes; the factory was not finished yet.

chapter seven

On June 7, 1939, Hitler made his second visit to the Volkswagen factory site. Riding in his open Mercedes limousine directly into the plant, he gazed up at the huge vaults whose concrete columns curved as if made of bent wood. In the months since he had laid the cornerstone, the great block of the power plant and wide, long assembly halls had risen rapidly along the Mittelland Canal.

The workers had held a topping-out ceremony to celebrate completion of the main structure. Traditionally, German workers set up a fir tree at the top of a building. For the VW factory, the tree was pruned into a topiary image of a Bug.

Now the building stretched along the canal, its vast roof punctuated with ranks of half-cylindrical clerestory windows and anchored at one end by the power plant, which looked like a virtual fort with smokestacks. Along the canal, twenty identical towers contained access stairs that marched along their sides with military regularity. At the entrance to each tower stood inspirational stone reliefs carved with imagery of heroic labor, symbolizing the various regions of Greater Germany—"Gaue" the Nazis called them, reviving a medieval word.

Inside the factory, Ferdinand Porsche and his assistants explained the machinery to Hitler's party. Soon, the huge green Hilo stamping machine, several stories high, would be in place. Between 1940 and 1976 it stamped out 30 million curved fenders whose shape became a kind of synecdoche of the Beetle itself. Before long, bicycles would wheel across this floor by the dozens, carrying workers from machine to machine.

Military production had already begun in the factory although no Beetles had yet been turned out. But to make the car a reality required funds, and in wartime Germany that meant military contracts.

The factory became a bazaar of various war production efforts as the Nazi leadership competed for control over armaments and raw materials. Himmler and Göring vied to create their own industrial empires around the SS and Luftwaffe respectively. Rational, central planning was supplanted by a jumble of internal politics. At the KdF factory, Ju-88 aircraft were repaired and their spare parts manufactured, and mines, shells, and bombs produced.

The Volkswagen itself had always been planned for easy adaptation to military uses. As early as 1937 Porsche's engineers turned out prototype versions mounted with machine guns. In May 1938 Franz Reimspiess sketched an early model of what became the Kübelwagen, or "bucket car," Porsche's Type 82; some 52,000 units of this scout vehicle were produced beginning in 1940. Many variants were produced for ambulance, radio, and other uses. In 1942 the Schwimmwagen, a vehicle with a watertight body and propeller, was introduced and some 14,000 were made.

The Kübelwagen was roughly analogous to the U.S. jeep, but did not have the four-wheel drive and hauling capacity of the American vehicle. But it worked hard in heat and cold, thanks to its lack of radiator, a feature that also meant no stray bullet could damage its cooling system. It was lighter than a U.S. jeep, making it less likely to get bogged down in sand and snow and easier to free if it did.

The German army was indifferent toward the Kübelwagen, but General Erwin Rommel told Porsche that it had saved his life: once the one in which he was riding ran across a mine without detonating it. The heavier vehicles that followed had blown up. In the desert war, where feint-and-parry was the rule, Rommel used the Kübel to bear plywood shells and create fake tanks. The Afrika Korps used a special model, with a mock tank body, to train its troops in the desert.

American troops sometimes drove captured Kübelwagen. The cartoonist Bill Mauldin described soldiers in Europe referring to them as (lowercase) "volkswagens." German army officers preferred big old-fashioned Horch staff cars, but the SS developed a special staff car version of the VW, made up of the rugged Kübelwagen chassis with the KdF body. With its fat, ribbed tires, this "Kommandantur Wagen" was the most sinister military doppelgänger of the Bug. Some of them had a fold-down map platform and, underneath the seat, fittings for a machine pistol and gasoline can, neatly marked with the SS twin lightning bolts. Hitler probably rode in one of them during the last days of the war when he visited the Eastern Front. It may have been the only time he rode in a Bug for other than ceremonial reasons.

Hitler enlisted Porsche in designing better tanks, including one with an electric drive of the sort Porsche had developed for World War I artillery and another, nicknamed "Ferdinand," to destroy other tanks. Had Hitler paid attention to the design principles behind the Volks-wagen—simplicity, durability, ease of repair—he would have understood the virtues of the Russian T-34 tank that was defeating his Panzers. Instead, he insisted on constant changes that culminated in the massive Mouse, Porsche model number 205, a 180-ton land ship, a virtual block-house on tracks, of which only two were built.

It is a vital part of the subsequent mythology that Americanized the Bug that no real civilian cars were ever produced under the Nazis. But a few did roll off the assembly line in June 1941. The numbers were very small, perhaps 600 depending on definitions. A 1941 factory photograph shows grim officials in front of the first civilian models, all destined for party bigwigs—Göring got one, as did Dr. Ley and others. Newsreels publicized the event to show the war had not diminished consumer production, but the images must only have reminded average Germans that owning such a car was further away than ever. After September 1939, despite extensive propaganda, few customers were foolish enough to sign up for the KdF savings plan. The cars could not hide their wartime origins: they had blackout headlamps—narrow slits that directed the light downward.

Porsche was determined to continue the Volkswagen project during the war. The disbelief he had expressed to Henry Ford about the arrival of war was succeeded by a belief it would be short. Porsche's concern above all was to keep the factory intact and working. He seems to have felt neither much enthusiasm nor much conscience about building weapons for Hitler, and his indifference to politics gradually developed into a wider callousness.

While vast camps of huts and barracks rose to house slave laborers, Porsche, in his country house surrounded by tulips, fought off interference from the Nazi warlords and clung jealously to the idea of the people's car.

At least once he took advantage of the dictator's indulgence of creative people and confronted Hitler directly: Did you not ask to build a people's car? The leader promised the civilian car program would continue. In fact, as the military tide turned against him, Hitler seemed to dwell more and more on his vast vision for the future and the vital place the Volkswagen held in it.

In September 1942, six months after his armies invaded the Soviet Union, Hitler traveled to his military headquarters in the Ukraine to be closer to the Russian front. His main headquarters in Berchtesgaden he called the Wolf's Lair, and another base in East Prussia the Wolf's Tail. But this camp near Vinnitsya he named "Werewolf."

Businessmen, generals, and hangers-on had assembled outside the Führer's simple cottage, dining at wooden tables, each with its vase of flowers.

Albert Speer recalled one visit. He could pick out dragons and castles in the big fleecy clouds towering in the sky. It was a lovely late-summer evening, green with tiny insects in the air, silent except for the occasional sound of an automobile far off.

Hitler spoke of the future until he was hoarse. "We will cross Egypt," he declared, "and after the Caucasus we will push on to Iraq and Iran. By the end of 1943 we will pitch our tents in Tehran and in Baghdad. We will advance south of the Caucasus and along the Caspian toward India. The British will run out of oil. Their empire will collapse.

"I will bring all the Germanic peoples together," he went on. Sixteen hundred years ago, the East Goths, the ancestors of his race, had paused here, settling for two centuries on the plain. Now Germans would return. He would build new towns and establish German farms. He would rid the area of Slavs, who would be made into forced laborers. The vast expanse of Russia would cry out to be filled and the new German settlers would move into quaint villages. He would extend the autobahns to the east to bind the empire together.

But he had questions for Albert Speer and Ferdinand Porsche. He wanted the German farmer who would settle the Ukraine to remain in touch with the fatherland. How far could the Volkswagen go in a day? How long would it take to drive to Berlin?

A steady fifty miles an hour, they promised. So it would be thirty hours from Kiev or Odessa to Berlin. Then, the Führer said, every village would need an inn for the travelers: a Gasthof zur Post. Perhaps he was thinking back to his own childhood when he lived in the inn called Gasthof zum Pommer at the village of Braunau on the edge of the Austro-Hungarian Empire.

Hitler's retinue listened to the leader's interminable monologue with growing tedium. Once, he recalled how as a prankish schoolboy he had sent his fellow students and teacher fleeing by releasing into the classroom a boxful of beetles.

He plotted out the landscape on the basis of the Volkswagen. The car was not just a delusion of propaganda, aimed at sedating the masses. It was a vital feature of Hitler's whole scheme, a literal measure for the landscape he planned to impose on the globe.

Hitler's vision of the towns planned for conquered Russia combined kitsch with brutality. Even Heinrich Himmler, the ruthless head of the SS, contributed his aesthetic ideas: in the windows of the houses and inns of the new villages, he suggested, there should be flower boxes.

And by 1942 Hitler had already built one such town and one inn—at KdF-Stadt, beside the Volkswagen factory. Despite the war, the building of the model city had progressed: managers lived in tall houses on streets lined with oak trees, just yards from barracks holding POWs and slave workers. There was a pond named after Schiller. But as Hitler spoke, his armies were about to suffer their worst defeats in North Africa and at Stalingrad. The factory he planned to devote solely to producing the Volkswagen was turning out military vehicles and weapons. Soon Hitler was demanding of Porsche and Speer not a people's car but miracle weapons, rockets and jet planes and hundred-ton tanks—"vengeance weapons" he called them. Soon all that would be left of the planned villages and the highways was the fantasy—and a few real, rolling Volkswagens.

The peoples Hitler wanted to sweep off the plains would build the cars for future settlers. The first Soviet POWs began to arrive in Fallersleben in 1941, and Poles, Slavs, dissidents, and convicts from military punishment battalions followed. Three quarters of the workers at the factory were slave laborers and were among the estimated 10–15 million that the Nazis employed during the war.

Rare documentary evidence of Hitler's direct involvement with slave labor in German factories lies in a letter he sent to Himmler authorizing the SS to take over part of the VW factory and bring in slave laborers:

[stamp—partially illegible] FILE NUMBER—Secret/250/20
[stamp] SECRET
[From:] DER FÜHRER

The completion and commissioning as well as the continued expansion of the foundries, especially the light metal foundry at the Volkswagen Werk, are to be accelerated with all means available.

I approve the proposal by Party Comrade Prof. Dr. Porsche and my representative Party Comrade Werlin to transfer the completion, expansion, and operation of these foundries to the Reichsführer-SS and Head of the German Police, who will provide the workforce from the concentration camps. The Reichsführer-SS is to take the responsibility for the implementation of these instructions in the shortest period of time. The factory must have started operations by the fall 1942 at the latest.

The necessary contingents for this purpose are to be made available immediately.

FÜHRER HEADQUARTERS, JANUARY 11, 1942

[*signed:*] ADOLF HITLER

TO: The Reichsführer-SS and Head of the German Police

TO: Party Comrade Prof. Dr. Porsche

TO: Party Comrade Representative Jakob Werlin*

Beginning in February 1942, the SS brought prisoners from the Neuengamme concentration camp, outside Hamburg, to complete construction of the light metal foundry. An encampment for the workers was built called Arbeitsdorf, or "work village." When the foundry was completed in September, the workers were returned to the camp. It was an early experiment in the use of concentration camp labor, and the SS was pleased with the result.

The completion of the light metal foundry meant the factory could manufacture additional military products—notably aluminum parts used primarily in aircraft.

In February 1942, Dr. Fritz Todt was killed in a plane crash and Albert Speer was named to succeed him and was soon promoted to chief of armaments and production. Speer found an economy ill suited for war, one that was still producing civilian amenities and inefficiently allocating scarce resources such as fuel, metals, and rubber. The biggest shortage was labor. German women were never mobilized for factory work, and the battles among the Nazi high command and interservice rivalries kept the economy from efficient production.

The use of slave labor was systematized when SS officer Fritz Sauckel was appointed in March 1942 as "grand plenipotentiary" in charge of

* Cited in documentation of class action suit, *Anna Snopczyk et al. v. Volkswagen AG*, Civil action No. 99-C-0472, U.S. District Court, Eastern District of Wisconsin, May 5, 1999, Exhibit E Certified Translation.

labor from the concentration camps, prisons, and occupied countries. Ordering up labor on demand became standard practice for almost all German industry, which used seven million slave and concentration camp laborers during the war.

Sauckel, who was often described as pig-eyed, was charged with a simple mission: the workers were "to be treated in such a way as to exploit them to the highest possible extent at the lowest conceivable degree of expenditure." Treatment ranged from reasonably feeding and even paying some workers to systematically working others to death. Sauckel established a hierarchy of workers according to nationality and race and shipped them to factories all over the country.

Volkswagen got its share. Not far from the factory, the huge Laagberg camp was set up to house workers and an SS guardhouse stood on the bridge crossing the canal to the factory.

While Bodo Lafferentz nominally headed up the VW business side and Porsche and Ley remained in charge of production, in 1941 Porsche's son-in-law, Anton Piëch, became factory manager. At the end of that year construction stopped on the civilian housing. A camp was completed for Italian laborers, sent by Mussolini's equivalent of the DAF, who assembled in a huge hall that remained a landmark for years. Named for the head of the group, Tulio Cianetti, it rose like a vast mad chalet, with the cogwheel logo of the DAF, the German Workers Front, and swastika symbol on the outside.

Soon the area around the factory became an encampment of low-roofed barracks, with antiaircraft guns and sandbags, and strange conical one-person bomb shelters.

The absurdities of the factory's operations were summed up by an emergency project for the fabrication of cast iron stoves for the Organization Todt, supplying the Russian front. The German army had expected to defeat Russia before winter and the troops were dispatched without winter clothing or equipment. By late autumn they found themselves bogged down in subzero temperatures.

In November the VW factory was ordered to carry out an industrial "shock action" and produce 100,000 cast iron stoves for the freezing soldiers. Casting such a crude product at the factory took no advantage of the high-pressure steel presses or light metal foundry. Stacked up for propaganda photographs to inspire the home front, the crude metal boxes looked like a parody of thousands of identical Beetles.

In early 1943, the factory built the Fi-103 flying bomb, better known as the V-1, the cruise missile that terrified Londoners called the "doodlebug." The V-1 caused more terror than damage when it began to strike London in June 1944, but Hitler put stock in it as a "vengeance weapon," and the British feared it. About three quarters of the 21,000 V-1s were built in the VW factory.

The Allies, tracking the V-1 program through photos and Resistance information, launched a bombing campaign called Crossbow aimed at the networks of labs and factories developing the buzz bomb and targeted the VW factory on April 8, 1944. Bombers returned again in June and August. Located in open country, the plant was an easy target. Plant officials had already established air raid shelters and protected key pieces of equipment with sandbags. In the camps around the factory a variety of underground shelters and small cast concrete shells sprang up.

During one raid, a British Lancaster bomber crashed into the factory. On another, a bomb fell close to the central dynamo but failed to explode. Had it detonated the whole factory would have been left without power.

As the war turned against the Germans in the east the supply of Soviet POWs dried up and the Nazis began enslaving laborers from the occupied countries. Most of them were simply called Ostarbeiter, "East workers," the all-purpose term for Poles, Ukrainians, and Russians pressed into a slave labor, as well as French, Dutch, Belgians, and others. These last were treated better, and some were actually paid. Another category of workers came from Auschwitz, Dachau, Bergen-Belsen, and other extermination camps. Many were women or children.

In March 1944, the Volkswagen managers applied to Sauckel for additional labor. Three hundred skilled Hungarian metalworkers were shipped from Auschwitz in the spring of 1944, 500 more in the fall, along with large numbers of women.

While the first groups had been comparatively well treated, these later ones were badly fed and housed. They ate little but turnip or potato soup with the occasional bit of horsemeat and customarily worked barefoot. They were frequently beaten and slept on straw mats.

Elly Gross was fifteen years old when on May 27, 1944, she was taken from her village in Romania and transported by cattle car to Auschwitz. At the end of August she was among a group personally selected by the infamous Josef Mengele to be sent to the VW factory where she worked fifteen hours a day. Lilly Klein was seized in occupied Hungary in March and sent to the factory in September. Both women later summarized their experiences for a class action suit seeking compensation for their treatment.

Some of the women arrived with young children; others gave birth while working in the factories. In 1943 Sauckel set up a system of maternity hospitals and child care centers, the Kinderheim, near the factories. Nazi policy regulated care for the workers and children according to ethnic background, and the children of Slavs and Poles were especially badly treated.

The VW Kinderheim was run by Dr. Hans Korbel and nurse Ellsa Schmidt, who frequently beat children and mothers. As the war went on and food grew scarce, the Kinderheim was increasingly neglected. The Kinderheim was in Ruhen, about five miles from the VW factory, and mothers could visit at most once a day. Underfed and covered with flies and roaches, babies were left to die, most within weeks. Their bodies were wrapped in toilet paper and piled together in cardboard boxes for burial.

One typical case was that of Anna Snopczyk, who was brought from her Polish village in 1944 and put to work fabricating bombs and shells. Suffering from an allergic reaction to substances used in the shells, she was relocated to a nearby farm that raised food for the factory. There she fell in love with a Polish man; their child, Josef, was born on February 11, 1945, and sent to the Ruhen Kinderheim where he died in April.*

* Snopczyk returned to Poland and never had another child; in 1999 she sued Volkswagen.

By the summer of 1944 Volkswagen, like many other German factories, was beginning to shift production to mines and caverns safe from bombing. Construction of the V-1 was moved to underground facilities near Tiercelet in western Germany and to caves in the Harz Mountains. These were part of the infamous Mittelwerke served by the Dora concentration camp, where Wernher von Braun's V-2 rockets were constructed. In the damp, cold subterranean factories, workers died even more rapidly.

As the Allied troops approached, many of the workers were sent back to concentration camps. The barracks for the laborers remained standing for years after the war and cemeteries scattered through the area held the bodies of thousands who had died—flat stones over the graves of Poles, iron crosses curling like cursive writing above the graves of Russians. Near the end of the war, someone carved on a tree on the Klieversberg the Russian phrase: "graveyard of Soviet youth."

chapter eight

tanks from the American Seventh Army arrived near the factory on the 9th of April, about a month and a half before Germany surrendered. Telephones and typewriters were smashed; slave laborers had taken their revenge where they could. Fearful ex-Nazi administrators and their former prisoners mingled in an economy where the prime currency was American cigarettes. The factory lay just fifteen miles from the Soviet front line, and while thousands of workers from the plant wanted to go home, thousands of refugees fleeing the Soviet army arrived seeking work.

Accusations and denials flew as denazification proceeded. In late June, Captain Clifford Byrum and fellow officers and men of the U.S. Army medical corps excavated and photographed the tiny corpses at the Kinderheim in Ruhen. An official of the facility had disobeyed an order to burn records as the Allied troops approached, throwing blank pages into a fire instead, and in 1946, Hans Korbel, the doctor in charge of the camp, and the nurse, Ellsa Schmidt, were tried by the British and found guilty of war crimes. Korbel was hanged the next year and Schmidt served several years in prison. After her release, a repentant and presumably enlightened woman, she was hired by VW as a social worker.

Fritz Sauckel, the master of the SS slave labor policy, was convicted at Nuremberg and hanged. Awaiting trial at Nuremberg, Robert Ley, the

DAF leader, had the temerity to write to Henry Ford, asking him for a job and telling him he had experience running automobile factories and organizations with large numbers of workers. Apparently without receiving an answer, he committed suicide.

Porsche was briefly detained by the Americans along with top engineers and scientists, including Albert Speer and Wernher von Braun, as part of a round-up code-named Dustbin. Porsche was released and returned to Austria. In October 1945, French officials invited him to Baden-Baden to discuss consulting work, then arrested him along with Anton Piëch, his son-in-law and the former factory manager. Charged with abusing prisoners of war, the pair were incarcerated in Dijon. While in jail, Porsche offered advice on the design of the 4CV, Renault's version of a French people's car. The French were eager to get their hands on the machinery of the plant as war reparations, but the Americans and British resisted. Ferry Porsche, back in Austria, sought consulting work to help pay a half-million-franc "bail" the French imposed on his father. Ferdinand Porsche and Piëch were finally released in 1947, in poor health.

The Potsdam agreements of July 1945 divided Germany among the occupying powers. The country was disarmed and the economy decentralized. Industry was limited to prewar levels. The Allied Control Council supervised zones of the country individually administered by the Soviet Union, the United States, Britain, and France. While the zones would eventually become economically unified, each nation was empowered to extract its share of reparations from its zone. The agreements anticipated the seizure of German industries as a source of compensation for the laborers and other victims. The Soviets dismantled the auto plants in their sector and shipped them back to Russia, further crippling German automobile makers, including Mercedes, BMW, and Auto Union, whose factories had been devastated by bombing. The Western allies concentrated on supporting the frail local economies.

KdF-Stadt and the factory fell under British control. In the summer of 1945 British army Major Ivan Hirst was home on leave when he received a telegram ordering him to Hannover and then to the plant in Lower Saxony.

Hirst was part of the elite Duke of Wellington's Regiment when he was evacuated from Dunkirk in 1940. Transferred to the Royal Electrical and Mechanical Engineers, he had returned to Europe in 1944 as head of a tank repair group in Brussels. With his beret and little mustache, his ascots and riding crop, Hirst was a parody of a British officer—a figure from an Evelyn Waugh novel. That the Bug should have found its savior in such a character seemed almost whimsical.

The British intended to use the factory to repair vehicles for the occupation forces. But Hirst's commanding officer, Colonel Michael McEvoy, had driven one of the VW prototypes before the war. Hirst and McEvoy had one of the few VW sedans still painted a military olive green and drove it to the headquarters of the British occupation authority. In September 1945, they persuaded the authority to place an order for 20,000 vehicles to serve local troops and officials and insured that at least for a four-year grace period the plant would not be dismantled. Without the order to resume production, the remaining machinery would likely have been disbursed among the occupying powers and the factory shut forever.

Hirst put refugees to work, scrounged steel and other rare materials, and repaired machinery. By 1946, 8,000 workers were turning out 1,000 vehicles a month.

It was clear that as part of denazification the town would need a new name and in August Wolfsburg was officially adopted to replace the formal KdF-Stadt and the informal "Autostadt" most residents used. The name was taken from Wolfsburg Castle. Built in the fourteenth century, it was once the home of the von Bartensleben family, whose crest bears the image of a wolf leaping over two sheaves of grain. The crest is carved above the entrances to the castle, the detail softened by the years. A less animated wolf would be hard to imagine: the animal appears to be lying on its side. There had been no wolves in the area for a century; the last one was shot in 1839.

A badge bearing the wolf and a stylized castle tower began to appear on the VW in 1951. The wolf stands atop the castle, looking back over his shoulder as if cornered, more indignant than fierce. The badge eventually became a coveted collectible for mischievous schoolboys on American streets.

Hirst helped recruit former Opel executive Heinrich Nordhoff to take over management of the company. Nordhoff had spent the war running an Opel truck factory in Brandenberg that eventually came under the direction of the SS. This degree of association with the Nazis disqualified him from management jobs in the U.S. sector, where he worked in a repair shop in Hamburg, under investigation for his past. In the British zone, the rules were more forgiving about ex-Nazi managers.

Nordhoff had never been impressed by the Volkswagen and was dubious about the whole enterprise, but he needed the job. He took over formally in January 1949 and would head VW until he died in 1968.

For a time it seemed that Volkswagen might be sold to another country. A much quoted formal engineering evaluation undertaken by the British military found little potential in the Volkswagen, declaring, "The vehicle does not meet the fundamental technical requirements of a motorcar. It is quite unattractive to the average motorcar buyer, is too ugly and too noisy."

The Wolfsburg plant was shown to several major car companies. It became part of the legend that Sir William Rootes, who with his brother Reggie had founded the British car dynasty known for Rootes and Humber cars, had told Hirst, "If you think you're going to build cars in this place, young man, you're a bloody fool." He predicted the operation would collapse in a couple of years.

Equally pessimistic was Ernie Breech, the former GM executive called in to tutor the young Henry Ford II, who was uneasily filling his grandfather's shoes. Breech warned the heir, "What we're being offered here isn't worth a damn."

Some British engineers did not totally condemn the Beetle, but they discerned that the noisy VW did not have the refinement the British public valued most in a vehicle. Nor would it have been easy for an English manufacturer to acknowledge even implicitly the superiority of German engineering by adopting one of its designs. Victorious Great Britain still saw itself as the country that produced the Spitfire, and its need for its own people's car would not be proven until more than a decade later with the creation of the Mini.

But there was another good reason for Western companies to avoid buying the operation. The factory lay only a few miles from the border with the Soviet occupation zone. Had they realized its value, the Soviets could easily have seized its equipment in the spring or summer of 1945 before the sectors were locked in—or even later. With tensions growing along the border in the late 1940s, British officers and plant managers kept cars ready for a quick getaway in case the Russians decided to invade.

But as the fractures of the Cold War took shape, there were growing second thoughts about the original Allied policy toward Germany. To many economists, it recalled the confiscatory edicts of the Versailles Treaty that had helped bring Hitler to power in the first place. To politicians, the need for a strong ally against the Soviet Union and East Germany required the restoration of both German industry and the rearmament of the country. The Bug became a weapon in the Cold War.

By 1946 tensions between the Soviet Union and the West were palpable. In May, Winston Churchill made his famous speech in Fulton, Missouri, describing the "Iron Curtain" that had fallen across Europe. On January 1, 1947, the U.S. and Great Britain linked their zones economically and the French joined two years later. The Soviets held out for their own plan to unify the zones. Fearing Stalin would take advantage of economic discontents in the West, George Marshall proposed his famous plan for the economic recovery of Europe in 1947.

By 1948, talks among the occupying powers concerning reunifying Germany broke down. In June, the West began drafting a new constitution and introduced a new currency. The Soviets responded with the Berlin Blockade, the West with the Berlin Airlift, and the borders were frozen for forty years.

Amidst such political uncertainty, economic recovery was slow. As late as 1949, a third of German workers were still unemployed. Nevertheless, in the fall of 1949, a special Beetle was introduced aimed at the export market, with better brakes, chrome accents, and other features, and despite the political climate, production expanded steadily. In the fall of 1950 Porsche visited the factory where his car rolled off assembly lines. Before he died in January of 1951, Porsche saw the autobahns filling with the cars he had dreamed of.

Hirst left VW in August 1949 when the company was set up as a trust run by the new West German government. As a departure gift, the staff presented him with a two-foot model of a green Beetle, not unlike the one Porsche had made for Hitler's birthday in 1938.

Hirst worked for a time at the Organization for Economic Cooperation and Development and in later years became a kind of hero to Bug buffs who visited him in England. "It was by no means a perfect car," Hirst would say again and again, "but in its time it was a damn good little car." Still dapper with clipped white mustache, ascot, and pipe, he became a familiar figure at the local pub near his home in Marsden, reminiscing until midnight.

Heinz Nordhoff, the man who took over the plant in 1949, was not unappreciative of the ironies of his situation—and the Bug's.

He had been among the group of Opel employees at the 1938 Berlin auto show whom the Führer dressed down for pushing their P4 model as a rival to the original KdF-Wagen. At the time Nordhoff had not been much impressed with Hitler's car. As he noted, "By one of the ironic jokes history is sometimes tempted to produce, it was the occupation powers who brought Hitler's dream into reality." Another was that after 1945, the British motor industry went into decline while West German industry led the economic miracle of the 1950s. The boom played a key role in Cold War propaganda as VW was celebrated as a benefit of the free enterprise system.

A 1954 *Reader's Digest* article waxed enthusiastic about "the town a little car had built." Nordhoff was described as "tanned and blue-eyed" and the town as a show window of Western progress. Nordhoff was revered in Wolfsburg as wise, noble, and aloof, much like Chancellor Konrad Adenauer. His command was almost total, but it was a little disturbing that the workers at Wolfsburg could abide any dictator, benevolent or not. The article touted the company's contributions and loans for housing and praised the $75,000 it had contributed to churches. Such stories built up the myth of Volkswagen as the industrial dynamo of the new free West Germany, lifted out of the ashes by the West to produce a democratic car for a democratic society.

But the Germans loved cars too well to be content with the Beetle for long. In the first years after the war, when the only cars available were a couple of Mercedes and BMW models and a motley assortment of "bubble cars," three-wheel Messerschmitts and odd Gogomobils, the Beetle looked like a real car. But its lack of heat, its noise, its tendency to grow holes in its floor near the battery, all became legendary national gripes. For many Germans the car was inextricably linked with the years of recovery and was something to be put behind them as soon as possible.

Today in a quiet courtyard of Wolfsburg sits a huge Henry Moore–like sculpture composed of two large ovals, like the pelvis of some dinosaur. The sculpture represents the double window of the original Bug, but in solid form: absence is rendered as presence, negative as positive. It is an emotional monument to the first years of recovery, when the negatives of hard work and hard times were turned at least in memory into the positives of building the company, society, economy, and individual lives.

The change from the double or split rear window to a single in 1953 wrought a subtle but important aesthetic change to the body of the Beetle. But for the people who made the car the change marked the end of the beginning for the company and the town.

In the first few years after the war refugees fled Poland and East Germany. They were hungry, desperate, and had recently been in fear of their lives. Many of them ended up working at Wolfsburg, and in just a few years, they found prosperity they could not have dreamed of.

A quite extraordinary sense of mission lay behind the set of rules laid down for the factory in 1947. This high-minded document—today it would be called a mission statement—declared that the factory should be regarded as held in trust for the German people and it forbade exploitation of the workers and any use of the facility for private gain.

The place was managed with a sense of common need founded on a legal status somewhere between private and public ownership. Posted in the factory were the "Ten Basic Laws of the Volkswagen Factory," including:

> The Volkswagen factory will doubtlessly become the property of the German people, whose interests will be safeguarded by popularly elected representatives. In the meantime, we are the trustees of the factory on behalf of the entire German people. Its buildings and facilities must be well maintained, and its machines, tools, and furnishings handled with care.
>
> All work at the plant is thus a service to the general well-being of the people. Exploitation of the workers or the pursuit of private interest cannot be tolerated.
>
> Together the workers and office employees make up a complete, democratically ruled production association. Each individual's participation in the management process is guaranteed by an exemplary company agreement with the works council.

In such an environment, and with high unemployment elsewhere, workers had little interest in joining the union. Managers granted benefits and wage increases automatically as production and sales soared, and the company supplied housing and built pools and skating rinks. Nordhoff persuaded museums to lend valuable paintings for art exhibitions in Wolfsburg. Major symphony orchestras gave concerts. The company's beneficent policies were a small-scale version of the social welfare net— with compensation for displaced persons, generous health care and vacation benefits—that the German federal government was building during the 1950s.

Peter Koller, Hitler's architect for the planned KdF city, had fought on the Russian front and spent part of the war in a Soviet POW camp.

When he returned to Wolfsburg to resume work on the town, he was a different man. Koller's new plans formulated in 1948 added churches. His Catholic church of St. Christopher (the patron saint of travelers was thought appropriate for an automobile city) was given a special blessing by the Pope. Its cornerstone came from the Colosseum, suggesting some relationship between the emperors who fed Christians to the lions and the Nazis who fed people to their war factories.

The original scheme that had put the town hall and other public buildings on the hill called the Klieversberg was abandoned. Instead they were placed in a city center closer to the canal. A modernist town hall went up in 1958, designed by Titus Taeschner, Koller's lieutenant—a stark white Bauhaus Rathaus. Alvar Aalto was commissioned to design a civic center, finished in 1962, and Hans Scharoun built a theater and concert hall on the hill above the town center.

But the city of the Cold War was a cold city, more socialist in style than capitalist. There was an odd sobriety and similarity about its buildings, as if they all came from a single hand, and the only cars visible on the streets were Bugs. Among the citizens, one of the most beloved landmarks was the Esso gas station.

What was amazing was how easily the basic town plan adapted to the shift in ideologies—a department store instead of a Nazi headquarters, in one case. To some, that was because the basic values of the plan were incontestable—nature and neighborliness. To others, however, the easy change suggested how much Adenauer's Germany still had in common with Hitler's.

In 1953, a photographer named Peter Keetman visited the factory just as the 500,000th car was produced. At the time Wolfsburg was turning out 6,000 cars a day and employing 20,000 workers. Volkswagen held 42 percent of the market, while the other German auto companies were still struggling, relying on one or two sturdy models or turning out offbeat little cars. Even more important to Germany was VW's value as a source of export revenue.

Keetman's photographs show some of the last cars with the original rear window. He worked in the mode of the German photographers of the Neue Sachlichkeit—the New Objectivity—like Otto Renger-Patzsch, and in the tradition of Charles Sheeler and Edward Weston, who photographed Fords and Ford plants. As Weston had zoomed in on the details of typewriters and Model T Fords during the 1920s, Keetman showed piles of gears and bumpers. Lovers of machine art would find a kindred spirit in his views of fenders nested like seashells, of piled cogwheels, of assemblages of cylinder casings with their cooling fins like sacramental cups. Few people are seen. Only a couple of his shots show the flaring sparks of silhouetted welders and workers mating body to chassis in the process called "the wedding." The viewer can't make sense of the process of assembly; there are few captions and the parts are treated as abstracts. The factory seems light and open, the work almost balletic. Keetman's was a high-minded and abstract aesthetic that concerned itself with neither the nobility of labor nor its exploitation.

A similar elegant and idealistic tone came through in the 1954 film about the factory called *Aus eigener Kraft*—"From Our Own Strength"—a celebration of the visual ballet of the assembly line that was the motion picture equivalent of Keetman's still photographs. There was even a novel celebrating the Volkswagen factory, Hans Moennich's *Die Autostadt*, published in 1953. Its cover bore a heroic image of the towering factory that would not have been out of place on a prewar KdF poster.

Many of the cars the film showed so magically taking shape inside the factory were shipped out of Germany, for the success of the company and the West German economy depended on income from exports.

The first efforts to sell abroad were not very promising. Promotional literature released in the U.K. in 1946 described the car as "the former Führer's people's car." No wonder that in the late 1940s a high percentage of the first Bugs imported to Great Britain were vandalized at the docks.

But by 1950 the car could be found in South Africa and South America. It was eventually accepted in the U.K. and beloved in France, where it acquired a reputation as the Coccinelle—or Beetle in French—to stand alongside the Renault 4CV and the smaller Renaults. Advertisements borrowed from the hovering, transparent Bugs shown in the early KdF promotions but gave them a Gallic twist. A Beetle flies magically among clouds above the tag line: *"Rêve d'hier, réalité d'aujourd'hui."* ("Yesterday's dream, today's reality.")

Thanks to booming overseas sales, in Germany the 1950s flashed by in a succession of two images. One is of ceremonies marking numerical milestones—50,000, 100,000, then half a million Bugs, with wreaths of flowers and parades and displays in stadiums.

In the spirit of the 1950s, the milestone car was usually raffled off to a worker. For the 500,000th car, completed in July 1953, a stadiumful of workers watched numbered Bugs drive in a tight circle, forming a kind of giant roulette wheel. A bouncing ball came to a stop by one of the numbered cars to designate the lucky winner.

The other set of images was of Beetles around the world, exported or built of parts or kits in factories in New Zealand and Australia, Nigeria and South Africa, Brazil and Mexico. The 500,000th car celebration featured a parade with music and costumes from each of the countries where the Bug was sold. Bagpipes and kilts followed skins and drums, and where banners with swastikas had once lined streets, the VW logo now flew.

The city and factory showcased democratic free enterprise as dramatically as Hitler once dreamed they would showcase his empire. They were symbols of the Adenauer prosperity. But around the world, the car was barely German anymore.

In each new country the Bug mutated slightly, attaching itself to local culture and traditions. Nowhere did this happen more dramatically than in America. There the Bug was wholly transformed. Thanks to the most famous advertising campaign in history, it returned to Europe in a wholly new guise, like a successful immigrant grown rich beyond his ancestor's wildest dreams.

despite the mythology that eventually grew up around its success, the Bug was a hard sell in the U.S. Two great salesmen failed. The first was Ben Pon, the slick Dutch exporter who had first ridden a Bug in 1939 (Porsche, wearing bedroom slippers, had given Pon a spin around Berlin in a prototype) and had led export sales in Holland as soon as the war was over. He tried to bring a few to the U.S. beginning in 1949 but found minimal interest. VW in Germany turned to Max Hoffman, the most successful U.S. importer of European cars, who had brought in Jaguar and Mercedes, and a legendary figure in the automotive world. If he could not sell this car who could?

Little was known about the car except that it had been Hitler's creation. Hoffman found it hard going even when he pressured dealers to take a Volkswagen for every hot sports car he was willing to sell them. Volkswagen hired a former Luftwaffe pilot named Will Van de Kamp to serve as evangelist, and he painstakingly crossed the country enlisting and inspiring dealers, like a Johnny Appleseed.

Volkswagen sales in the U.S. did not begin their dramatic rise until the second half of the 1950s. From 1953 to 1959, they grew from 2,000 to 150,000—through the efforts of avid local dealers and favorable publicity in magazines from *Popular Mechanics* to *Reader's Digest*—but without national advertising. And this was a car that until 1955 did not even have modern turn signal lamps, but mechanical "semaphores."

The Bug was discovered by word of mouth, like a good restaurant. It was a kind of found item, without history. Like any immigrant, it could reinvent itself in America, like Raymond Loewy or Louis B. Mayer or a Mafia boss who began as a Sicilian peasant boy. To ignore, elide, exaggerate, or simply reinvent one's past was the American way and that reinvention would, in part, come from advertising. While Detroit under the sway of Harley Earl tapped into the American values of power, speed, flight, and luxury, VW emphasized another set of American values, simplicity, durability, and economy.

VW decided to advertise to maintain its market position when Detroit was about to launch its compact cars in response to the Beetle. In 1959 the company hired the advertising firm of Doyle Dane Bernbach, and their campaign would become the most famous ever—and the qualities of the ads were those of the car: modest, tough, modern—and not ones usually associated with Detroit.

Within a couple of years everyone knew about the ads. The Starch survey found that an astonishing 70 percent of magazine readers paid attention to the advertising copy. Within five years, the ad campaign the agency created was already considered a classic; by the end of the century it was repeatedly voted the consensus best ever and William Bernbach, the art director Helmut Krone, and the copy writer Julian Koenig were venerated on Madison Avenue.

VW and the advertising agency with which its image would be inextricably linked found each other the way Americans found the Beetle in the early 1950s.

Helmut Krone, in fact, had been one of the 156 Americans who bought a Bug in 1950, nearly a decade before the agency went to work for VW. The classic "early adopter," unintimidated by being the first to buy, Krone chose products on the basis of the same utilitarian virtues the VW ads offered: low price, durability, economy of operation, high resale value lent by unchanging shape. He was in a sense a model of the ideal buyer—not the person who bought the cheapest car, but the sensible, well-educated person who bought a car no more expensive than he needed.

Carl Hahn, the young German who had been thrown into the job of directing Volkswagen in America, discovered the agency almost by chance. Hahn was Heinz Nordhoff's protégé. A bright young man whose father had worked for Auto Union before the war, he had impressed Nordhoff by writing him an unsolicited memo of advice. In 1959, Nordhoff dispatched Hahn to handle the important U.S. market. Hahn looked at the work of dozens of larger, more traditional agencies from the WASP and martini world of Madison Avenue, circa 1959, and was appalled by all of them. The ads these shops submitted seemed interchangeable, the proposals simply replaced a tube of toothpaste or box of detergent with his car.

DDB had been hired to handle ads for the opening of a dealership in Queensboro, New York, by VW executive Arthur Stanton, on the strength of the agency's irreverent work for Ohrbach's department store, for which the most famous line was "SNIAGRAB"—"bargains" spelled backward.

When Hahn attended a breakfast at the Plaza Hotel to discuss the campaign Stanton had commissioned, it turned into a multihour consultation about a single sentence in the ad. In contrast to the formulaic treatments he had seen before, this agency's attention to a few words impressed him.

One of DDB's most memorable ads showed American Indians and Chinese happily munching on loaves of rye bread over the slogan, "You don't have to be Jewish to love Levy's."

That the agency's founder was Jewish was also a factor, as were many of its clients, from Levy's to El Al Airlines to Ohrbach's. To hire Jews to sell Hitler's car brilliantly disarmed its greatest liability.

Bernbach lived not in the WASP confines of Greenwich, Connecticut, or Westchester County where most Madison Avenue CEOs lived, but in Bay Ridge, Brooklyn. He had grown up in the Bronx and worked as a copywriter at Grey Advertising, one of the few Jewish firms in the business. In 1949, he left to form his own firm with Ned Doyle and Maxwell Dane. But it was not the agency's religious background that made it Jewish. It was its style and the tone of its humor.

As Stephen Fox noted in *The Mirror Makers: A History of American Advertising and Its Creators:*

The atmosphere at DDB partook more of the Art Students League and the Seventh Avenue garment district than of the Ivy League or Greenwich. Most important, DDB produced ads that were unabashedly, recognizably Jewish in style and attitude. Their clients were all little guys, plucky struggling newcomers standing up to the bigger privileged competition, using their wits and humor to avoid being squashed. The funny ads provoke smiles in the characteristically Jewish fashion. Of self-deprecation from strength.

Self-deprecation from strength—it was a long way from strength through joy. It was closer to the battle of the little guy against the bullies—Charlie Chaplin's tramp, or Mickey Mouse—Disney's "great unlicked"—or Popeye or Kilroy.

Fox notes "the 'klitchik' at the end of the ads, the punch line derived from traditional Jewish humor." Examples were: "In one era and out the other," a swipe at American car styles that quickly became obsolete, and "How long can we keep handing you this line?" beside a silhouette of the Bug.

This was a novel approach to selling a German car fifteen years after the war. "Hell yes," it was a problem, admitted George Lois, who worked on the account before setting out on his own distinguished career. "In 1959 the car still reminded you of the ovens."

That such an approach, that such people should make Hitler's car a success was one of the most exquisite ironies of the Bug's story.

One WASP ad man compared the voice of DDB's ads to a stand-up comedian transplanted from the Borscht Belt. But the artist Bernbach admired was jazz musician Thelonious Monk. Bernbach was a bit mystical about creativity, often citing with approval Monk's line, "Sometimes I play music I don't hear myself."

"Creative" was the word most often used to describe their approach. But creativity was not what was lacking in the copy and paintings that evoked Pegasus and rocket ships in Detroit's ads. The key difference was tone, the way the words and the images worked together, in the same key and on equal footing. What was creative was the coordination of words and images in an era when most copywriters and art directors worked separately. Bernbach brought in modern ideas of the visual working with verbal when he hired art director Robert Gage, who had absorbed ideas from influential graphic designers Alexy Brodovich and Paul Rand, who, in turn, had learned from the Bauhaus and Swiss modernists.

"An honest car honestly presented," was the pitch.

It was a disingenuous proposition that may have taken its keynote from a *Popular Mechanics* story of 1956 whose theme was that the Bug was "an honest piece of machinery"—a sentiment that could appeal not just to the hard-nosed-garage-do-it-yourselfer who read the magazine but to the modernist-influenced intelligentsia. Not by chance, the author of the story, Arthur Railton, later became head of public relations for VW of America.

An honest car honestly presented was a noble theme, to be sure, but as Bernbach realized, the real trick was "not to tell the truth but to tell it artfully enough so that people would believe you were doing it."

The first pitches were basic—the car was cheap and durable and didn't go out of style. It was well made and inexpensive to maintain and run.

But as important as this "honest" presentation was the basic strategy from which it grew: the first goal DDB came up with was to Americanize the product. This strategy evolved after the agency sent a delegation to the factory. Like Porsche at River Rouge, like Peter Keetman in Wolfsburg, the ad men discovered inspiration in the processes and the workers. "We immersed ourselves in the process of making the car," Bernbach recalled of the delegation's three weeks in Wolfsburg. There is no evidence the ad men visited any other maker's factory for comparison—Did they know anything about auto factories? Did they have any reason to think Chevrolet or Ford or Studebaker did not have as many quality inspectors or coats of paint as VW?—but the very fact they went at all was unusual.

The first fruits of the trip were seen in the famous "Lemon" ad. It showed an apparently perfect Beetle with the words: "This Volkswagen missed the boat. The chrome strip on the glove compartment is blemished and must be replaced." Lemons never made it, the ad boasted. The number of quality inspectors was another boast: 3,389 of them. One was even identified by name: Kurt Kroner.

The campaign began with the famous ad headlined "Small Is Beautiful": "These strange little cars with their Beetle shapes were almost unknown. All they had to recommend them was 32 miles to the gallon (regular gas, regular driving), an aluminum air cooled rear engine that would go 70 mph all day without strain, sensible size for family and a sensible price tag too. Beetles multiply; so do Volkswagens."

Another recurrent theme of the ads was the number of invisible improvements in the Bug—better engine, new convenience features. While others wasted energy on outward appearances, Volkswagen bragged, it concentrated on changing the things that mattered. It was an argument for the mind over the eye. It flattered the customer with the implication he was smart enough to appreciate these invisible technical changes—rather than the superficial exterior changes Detroit offered each fall—and it suggested he was a person who appreciated quality even in simple things.

The ads systematically took on each genre of American auto advertising. "Our engine has no Super Turbo Injector" read one ad. But gradually other themes were added—social ones. "Keeping up with the Kremplers" showed all the other goods you could buy with the money saved by buying a Beetle—and struck at the "status symbol" structure of car marketing. There were subtle put-downs of Detroit. "Don't forget antifreeze," was one. "Presented by Volkswagen dealers as a public service to people who don't own cars with air cooled engines." But there were none of the slogans about "German engineering" that Volkswagen would later employ. And no references to Ferdinand Porsche, a name now associated with the sports car company.

The ads cleverly manipulated history. "Ten years ago, the first were imported . . ." the "Think Small" ad began, clearly drawing the starting line for the Bug in the postwar period. The ads of course did not say the Beetle's basic design was already ten years old when it arrived in the U.S., or where it was imported from, or who had designed and built it. The word German was never used, although funny-sounding, harmless

German names or words were. The ads ignored the car's greatest vulnerability, its Nazi origins, and turned its second greatest one, its age, into its
biggest asset.

The car was presented as timeless, so timeless it was almost natural, like an egg. Its style appeared outside of time in an America where
the automobile model changes were an annual event. While some ads
disarmingly referred to the car as ugly, others suggested it was perfect.
One ad claimed that the company had gone to "a famous Italian
designer" and asked him to improve the Beetle. He thought long and
hard, the story went, then made a single suggestion: enlarge the rear
window. The ad copy was apparently based on the fact that VW had
hired Pininfarina, the great designer of Ferraris, as a design consultant
in 1949.

The most revealing time game involved the "1949 auto show" ad of
1959, one of the first of the ads. It showed a Bug with a Cadillac, a Hudson,
a Packard, and a "coming and going" Raymond Loewy Studebaker. "Where
are they now?" ran the headline.

The concept worked so well that it was remade in 1969 as a television
spot. In addition to suggesting that the Bug had been new in 1949, it
implied the American models were obsolete. But today, those cars are
collected and venerated as classics.

The DDB team claimed to want the tone of "one person simply talking to
another," not yelling or cajoling. What resulted was a little different: the
editorial we, of *The New Yorker*'s "Talk of the Town" section.*

This wry tone succeeded because it was right for the time and the
target audience. It was cool, in Marshall McLuhan's soon to be famous
distinction, not hot like the shrill ads of other carmakers. Volkswagen
was not aiming for the average customer but for an upper-class customer
who appreciated the Bug's economy and simplicity in contrast to that of

---

* The company showed it understood this by putting together a special promotional book-
let of cartoons of cars from *The New Yorker,* called "Small Is Beautiful," which was distrib-
uted at dealers.

American cars. There was a touch of WASP puritanism in this appeal as well as a dose of reverse snobbism.

Visually, the ads used simple means to suggest the idea was being shown directly and honestly. They used black and white photography at a time when most other automobile ads used colored paintings. There was lots of white space, with the images tucked off in corners. The most famous ad was the empty page that carried only the line, "We don't have anything new to show this year." The sans serif type signaled modern, European, neutral, Swiss. The copy was intentionally laid out with ragged margins and widows and sentence fragments. Half lines and half sentences formed an impression Krone called "Gertrude Steiny."

The ads grew naturally out of the car's virtues while ignoring its vices; they expressed the impact the car had been having on the marketplace for a decade. But they also grew out of American values: simple shape, simple technology.

The Beetle did not look old—or new. It simply spoke a language different from the language of American automobile design, with its styling cues and brand themes—the sports chat any red-blooded American could speak.

In 1959, the time was right for the new approach to automobile advertising and the reasons can be summed up in two words: Sputnik and beatnik. The Soviet launch of Sputnik in 1957 symbolically humbled America's overblown technology as embodied in its popular rocket-finned cars. And "beatnik"—a word introduced by the newspaper columnist Herb Caen that popularized the impact of the literary "beats"—marked the first stirrings of a counterculture revolt against the mainstream ethos of consumption, the culture of the new and improved product, and the annual model change.

For many Americans, German technology still meant precise, prestige products like Leica cameras and Mercedes-Benz automobiles, Braun electronic equipment and razors. People also associated it with formidable weapons—Tiger tanks and Messerschmitt fighters. (That Messerschmitt, which had produced the world's first jet fighter, should be reduced in the 1950s to turning out three-wheeled bubble cars had not entirely diminished this prestige.) Meanwhile, Hollywood discovered the Beetle's cousin, the Porsche, the car in which James Dean raced and in which he was killed.

If one needed further reasons to trust in German engineering, Sputnik offered them. As Americans armed themselves for the Cold War, German engineers and scientists joined this effort. Wernher von Braun, who like Ferdinand Porsche had been briefly detained by Allied troops, led the team that finally got an American satellite into orbit to match Sputnik. The Vostok space capsule that in 1961 carried the first Russian into space, Yuri Gagarin, was often described as "about the size of a Volkswagen Beetle."

Since World War II it had been gospel in Detroit that the real people's car, the inexpensive model, the first car, was a used car. There was no need for small inexpensive cars in America when there were so many large inexpensive used ones. Ford executive Ernie Breech made fun of VW in the mid-1950s, saying "what they sell in a year we produce in a day." But as the sales of Volkswagens and other European cars and American economy models like Rambler and Nash increased, Detroit relented. It introduced small cars, although auto executives could still not bring themselves to call them that. They were "compact" or "economy" cars: Chevrolet Corvair, Plymouth Valiant, and Ford Falcon. But Detroit failed to see that what drew American buyers to smaller cars was not just low prices but their practicality and sportiness. Compared to European models of any size and price, American ones rode more softly but consequently handled less crisply in turns. Their engines were powerful but their brakes were weak.

In October 1958, *Life* magazine ran an article called "Beetle Go Home," which reported the surge in sales of foreign cars and the imminent release of Detroit's response, the "compacts." The article observed that the European small car was "no more a fad than the growing U.S.

taste for foreign foods, clothes, and decor." The common people's car was not being purchased by common people, it noted. The VW buyer was not the working-class person "but rather the high income architect or doctor."

In the same issue of *Life* were ads for the Edsel, introduced the previous fall—"Dramatic Edsel Styling Leads the Way!" the copy screamed—and for Chrysler: "Step up to the mighty Chrysler. Ease behind the wheel. Lounge and feel the deep carpeting underfoot. Survey the eye-catching world about you. Could any throne be more commanding?

"Push a button and head for the open road. Never have you dominated motion, space and time so completely." The verbal chrome was another sign that the VW's enemy was killing itself off with excess.

The first American compacts were introduced in simple, stripped-down form, but dealers sold more of the models with expensive options added, suggesting that what attracted buyers was style and sportiness, not economy. The Chevrolet Corvair was the most innovative of the compacts, with a rear air-cooled engine like the Beetle, but it sold slowly. It leaked oil and had an ineffectual heater and relatively poor gas mileage for an economy model. The model that succeeded it was the Monza, a higher-powered coupé, a personal sports car owned by the likes of Priscilla Presley.

Another compact was the Ford Falcon, a simple, even dull-looking car. Legend had it that Robert McNamara, the Ford executive who later ran the Pentagon for John F. Kennedy, drew up the specifications for the Falcon in church, on the back of a hymn list.

Despised by true "car men," its underpinnings would nevertheless serve as the basis for the Mustang, the personal sports car that made Lee Iacocca a legend. "The car you design yourself," was Ford's slogan when the Mustang premiered at the New York World's Fair in 1964. Buyers typically ordered enough options to double the car's base price.

When the compact wave began in the late 1950s, many Americans simply wanted something different, not only or even primarily something that would save them money. They liked the idea of a simple small car, but were more interested in owning a personal car than a people's car. But by 1959 it was clear that American design and advertising excesses were alienating a growing portion of the public. The painstakingly highlighted gouaches and oils of American auto advertising were, like the cars, masterpieces of their type. But the buying public was ready for something that was as democratic and universal as blue jeans or the jeep.

In 1958, the Beetle owner was still comparatively rare, but other European influences were arriving in the college towns where VWs and MGs and Renaults had first showed up. People were buying Breuer chairs and Danish modern coffee tables and reading Günter Grass's novel *The Tin Drum,* in which little Oskar, the hero, beats his way through the Nazi years.

Imagine the stereotypical Beetle buyer of the 1950s, relaxing in his Eames lounger, the African mask on the wall, the natural driftwood sculpture on the mantel, listening to Coltrane and Monk on the hi-fi, and talking about the latest Bergman or Fellini movie. But if the Bug first found favor in college towns it soon tapped into wider changes.

Vance Packard's *The Hidden Persuaders,* which stayed on best-seller lists in 1957 for eighteen weeks, criticized the manipulations of marketing and advertising. It especially attacked the school of "motivational research," which Detroit had embraced in order to appeal to the Freudian drives of American males. Ernest Dichter, the prophet of the movement, combined Freudian analysis with status marketing. One of his famous assertions was that men thought of a convertible as a mistress but of a sedan as a wife. The automobile, he declared, is "one of the most perfect psychological devices ever devised for sublimating our subconscious wish to kill or be killed." He advised manufacturers to overcome a certain inherent puritanism in the American mind. Ads, he argued, should persuade the buyer to let loose his desires by granting moral permission to have fun without guilt.

But even Dichter saw trouble ahead for the status system. He warned his clients in November 1958 that the symbols of status had changed. It had become ostentatious to have too big a car with too much chrome. Now people wanted "the car that is more honest, more real, a truer car."

High-status executives, architects, or television directors, he said, were turning in increasing numbers to more compact cars in their "search for visibly distinctive status symbols." The rank-and-file buyer confronted with the search for even bigger cars and symbols of higher status was confused.

In 1960 Vance Packard next attacked planned obsolescence in *The Status Seekers*. He argued that in the late 1950s the power of established status symbols had declined because prosperity had brought them within reach of anyone. Hollywood stars were buying European sports cars to show off; small was becoming expensive. In this sense, Packard proclaimed the end of the Packard, once the ne plus ultra of luxury cars, which had merged with humble Studebaker in 1955. Class had come to the masses. "Populuxe," critic Thomas Hine would name it. In a boom economy and a democratic system, anyone could aspire upward: the owner of the company driving a brand-new Caddy could find himself sitting at a stoplight beside the union boss in the identical car.

DDB's VW advertising campaign undercut the carefully constructed status language Alfred Sloan had pioneered and offered an implicit critique of novelty, indeed of an American society where new and improved was the most popular slogan. "If you want to show you've gotten somewhere," went one ad line, "get a big beautiful chariot, but if you simply want to get somewhere get a bug."

The American automobile industry had been built on the policy of trading up the line of models. General Motors's job was not just to manufacture cars but to manufacture discontent. VW took direct aim at that system. They "manufactured discontent" with the system of "manufacturing discontent."

To burnish its image as modern and international, Volkswagen standardized its dealerships in International Style buildings whose clean appearance suggested reliability and dependable maintenance. Its model was the Howard Johnson restaurant, a reassuring presence Americans found near the new interstate highways.

Using simple geometric shapes along with the "lollipop" sign, its round blue VW logo set atop a pole, the crisp white dealerships often were decorated with the image of "Mr. Bubble Man," the cartoon service technician with his neat bow tie, a distant cousin of such commercial cartoon characters as Speedy Alka-Seltzer or Reddy Kilowatt.

This clean efficient image, in contrast to the often scruffy dealer operations of Renault or Fiat, was one reason that VW—alone among the imports—was able to retain a strong position after the introduction in 1960 of American compact cars, which were maintained and supported by vast dealer networks.

Gensinger Volkswagen in Clifton, New Jersey, the oldest dealership in continuous operation, was a light brick building with a Greek Ortho-dox Church towering above it and a highway in front—an apt location. Its high-pitched roof was punctuated by twin chimneylike structures topped by blue VW lollipop signs that appeared to be floating like per-fectly round blue puffs of smoke.

Competing on the high walls inside were mounted swordfish, dol-phins, marlins, and sharks, and great heraldic seals, apparently German in origin, bearing griffins, bears, lions, and wolves. Two huge blue VW logos loomed beneath the sound-absorbent panels on the ceiling. Whether the Beetle was modern or not, its house was.

As important as Sputnik was beatnik: the Bug attached itself not only to the new modern culture but the emerging new counterculture. And in the view of Thomas Frank, author of *The Conquest of Cool,* and an observer alert to any sign of the commercial coopting of the countercul-ture, the Bug never sold out. It kept its legitimacy right into the 1960s.

"For many countercultural participants, the Volkswagen seemed an antithesis to the tail-finned monsters from Detroit," Frank wrote, "a symbolic rebuke of the product that had become a symbol both of the mass society's triumph and of its grotesque excesses. The VW was the anticar, the automotive signifier of the uprising against the cultural establishment."

It spoke a different language, but it still spoke in tones of taste and class: you had more taste if you bought a Beetle than if you bought a garish American car. It was an almost philosophical argument about truths beyond appearances. "Ugly is only skin deep" ran one ad line.

This was understood in a shrewdly reasoned tribute to the Bug's ad copy and its writer Julian Koenig by *Ad Week* critic Bob Garfield in 1999. "To be amused by Koenig's copy was to be flattered by it," Garfield argued. "The car that presented itself as the antidote to conspicuous consumption was itself the badge product for those who fancied themselves a cut above, or at least invulnerable to, the tacky blandishments of the hidden persuaders.

" 'Think small' was thinking quite big, actually. The rounded fenders were, in effect, the biggest tail fins of all, for what Volkswagen sold with its seductive, disarming candor was nothing more lofty than conspicuously inconspicuous consumption. Beetle ownership allowed you to show off that you didn't need to show off."

Advertising helped Americanize the Bug. But the car's final success was not just to become American but to transcend nationality. By 1960 it seemed to have been around so long that it no longer came from a particular country, or even a particular company. And the origins of that company had been downplayed to the point of a whisper.

"Hitlermobile" had become a mild joke. When journalist Dan Greenburg produced a "Snobs' Guide to Status Cars" for the July 1964 *Playboy*, his rules for "how to own a Volkswagen" included this line: "If you are Jewish and somebody should ask you what kind of a car you drive, say: 'A VW, and I know, but it's a helluva solid little piece of machinery.' "

Although the ad men at DDB thought they had abandoned their original idea of "Americanizing the product," in fact they had done just that—by another route. The values of the ads were American values—frugality, integrity of design, honesty of shape, economy, lack of pretense—not the values of the Cadillac—but American values nonetheless, with a long tradition. The Bug was the underdog, the spiritual sibling of Chaplin's tramp and the heir to the Model T. In ads embracing American values, the car showed up like the immigrant who becomes more patriotic than the native.

The ads also emphasized features that transcended its origins and style and nationality. After the first Moon landing in 1969, VW ran an ad showing the landing module and the line, "It's ugly but it gets you there." The analogy touched on what appealed to Americans about the Bug: its no-nonsense quality, the pure undisguised shapes, the can-do, make-it-work, jury-rigged quality of the machinery.

In the matter of its water worthiness, the Bug came close to folklore. That the car was so tightly built that one often had to lower a window to get a door to seal against the pressure had been noticed even in the KdF models. Of course such pressure also suggested the small interior volume of the car. In 1962, an Italian eccentric put a propeller on a Beetle and sailed it across the Straits of Messina between Sicily and the tip of the Italian boot. For a 1962 article in *Sports Illustrated* magazine a Bug was dropped into a lake at Homosassa Springs, Florida, where it floated for nearly half an hour.

The ads created a kind of technological folklore about features of the car's design. "No more functional shape could be designed," the ads intoned. The shape of the Bug was not an artifact, the message implied, but a natural feature.

chapter ten

by 1960, West Germany at last reached the per capita rates of automobile ownership the U.S. had achieved in 1930. The same year, VW became a joint stock company. The state of Lower Saxony and the West German federal government each held 20 percent of the shares, but ordinary Germans also bought stock, notably workers in Wolfsburg. The company would now be controlled by the often divergent interests of politicians, managers, and labor leaders and by stockholders who watched share price. The old ideal of being held in trust for the German people was forgotten.

It was a symbolic turning point, but the Bug itself had already approached the pinnacle of sales and popularity in West Germany and was being perceived by shrewd observers such as critic Reyner Banham as past its prime. People bought the Beetle for symbolic reasons, he argued. "Its overwhelming virtue in the eyes of men of liberal conscience was that in a world of automotive flux, its appearance remained constant, and in a period when cars grew larger year by year, it remained the same size. In other words, it was a symbol of protest against standards of Detroit, the mass media and the Pop Arts."

Banham declared its technology too old and its "timelessness" a myth. "News that the Volkswagen is unchanged will be better when it stops," Banham wrote in the *New Statesman* in October 1960. The Bug's reputation, he charged, had its roots in "a combination of Platonic aesthetics with ignorance of the nature of technology, summed up in the slogan 'a good design is forever.'"

As for Volkswagen's claims of constant improvement under the skin, Banham argued that "if you averaged out the technological improvements made in the VW over 25 years they were fewer than in most cars that have offered annual styling changes as well."

But during the 1960s, the Beetle was greatly transformed for the American market. The promise to change the car invisibly, under the skin, that the ads had made for so long, had at last been kept. The process began in the early 1950s. The changes included improved lighting, turn signals, and brakes, but above all, larger and larger engines. In fact Heinz Nordhoff claimed that the 1954 Beetle shared not "a single bolt or screw" with the 1953 model.

The most dramatic changes were the results of the first safety regulations of 1967. The instruments and dash were padded, the steering wheel and door locks changed, and the seats upgraded to what were called "sarcophagus seats." The engine would get fuel injection; a modified automatic transmission was available.

But the car's basic technology was years behind its European and Japanese competition. In 1968, *Road and Track* magazine marveled that "the Beetle is a genuine anachronism, a genuinely updated old car—old enough now to almost qualify as a modern replica of itself." And still it sold. In 1968, 5 percent of all new cars sold in the U.S. were Bugs and it was the best-selling car in Southern California. While sales had peaked in Germany in the early 1960s, they would continue to rise in the U.S. for several years.

In the United States, the Bug ideal would undergo many mutations. In California, especially, that hothouse of automotive ideas and fantasies, the Bug would make itself at home, as so many immigrants do, and tap into the existing cultures, reshaping them in the process.

In hands as different as those of Walt Disney and Charles Manson, the bright ideal of universal automobility mutated into expressions of different California ideals. From its shell was born the "Cal Look," the lowered, souped-up racing Beetle. From its bones—the chassis and engine—evolved the dune buggy.

The Cal Look Beetle was to the original people's car what Ed "Big Daddy" Roth's Rat Fink figure—the snarling hot rodder's mascot—was to Mickey Mouse. Aggressive, angry, and high-powered, the Cal Look was a customized low-slung vehicle to which had been applied variants of the chopping and channeling more commonly inflicted on late 1940s Mercurys and old Model A Fords. Its engine was modified for additional power, its body lowered and extended.

The dune buggy was an even more radical transformation. It tossed away the Bug's basic body and turned the Volkswagen engine, frame, and suspension into the basis of vehicles at home on the beach or desert. At Glamis Desert and Pismo Beach evolved a special variant, the rail, built on a Bug chassis that was thinned and extended into a form so specialized for the steep dunes of beach and desert and special races that the car resembled a bird of the same name.

A boat builder named Bruce Meyers fabricated a sleek fiberglass shell to top a Bug chassis. The result was the Meyers Manx, a 1960s symbol of cool. Elvis Presley drove one in the movies and Steve McQueen traded his more familiar Mustangs and racers for a Manx in *The Thomas Crown Affair*.

Hot rod culture had begun on the beaches and on the dry lakes and deserts in the 1920s and 1930s with candy apple red paint jobs, chopped and channeled, lowered and lengthened cars. It was an edgy culture of controlled rebellion and stylized adventure, where the flying eyeballs of Moon auto parts ogled grinning monsters like Ed Roth's Rat Finks.

Reflected in the decoration of surfboards and the paintings of local artists such as Robert Williams, Ed Roth's house artist, who also sold through art galleries, California car culture intersected with the worlds of art and film. *Kustom Kar Kommandos*—the title of Kenneth Anger's legendary 1965 cult film on the subject—summed up the car culture's combination of engineering, cartoon, and kitsch. With the Cal Look and souped-up engines, the Beetle moved with surprising ease beyond cute and practical to aggressive and exciting. Beetles were outfitted with Rolls-Royce grilles or tail fins, parodying big, rich cars. Some convertibles with surfboards in the back seat became sports cars. Soon there were bad Beetles and mad Beetles, as inflected with adolescent energy and cartoon craziness as the cars and imagery of California customizers George Barris or Von Dutch—vehicles for the wild surf Nazis who hung out on the beaches. Nazi regalia, after all, was one of the best ways to shock people, which was the point.

On March 13, 1969, Walt Disney's *The Love Bug* premiered in theaters across America and became the most successful film of a year that also brought *Easy Rider* and its counterculture visions of apocalypse on the open road. *The Love Bug*'s box office status was due in large part to its PG rating and the fact that the great baby boom had peaked in 1960, providing Disney with a bumper crop of nine-year-olds.

The film's premise was that Herbie the car was alive. Disney took the shell of the car and filled it with cartoon concepts and trick mechanicals.

There was nothing new about the red and blue stripes on Herbie's body, no special reason behind the number painted on his side. Bill Walsh, the producer and writer, took the number 53 from Don Drysdale, the Dodger pitcher. Herbie the Love Bug was a spiritual descendant of Fred MacMurray's flying Model A in the "Flubber" films, *The Absent-Minded Professor* and *Son of Flubber*.

For adults, the car's persona evoked, if distantly and without depth, the little man characters familiar from Chaplin or Buster Keaton films. The very name Herbie suggested a Walter Mitty–like persona.

For Disney, *The Love Bug* offered a sort of vaccination against counterculture infection. Nineteen sixty-nine was a year of violence and ugliness. The day the film opened, the news was full of the costs of Vietnam: 432 Americans had died in the most recent Vietcong offensive and Defense Secretary Melvin Laird was asking for increased spending on the war. Interior Secretary Walter Hickel met with Florida governor Claude Kirk to discuss the problem of alligator attacks on humans. *The Love Bug* offered an escape.

But the film was also Disney's bid for hipness. "Look who's running with the fast crowd!" the theater posters proclaimed. *The Love Bug,* in the familiar Disney strategy, soon became one of the company's franchise properties, like Peter Pan or Goofy. *Herbie Rides Again* arrived in 1974 and three years later *Herbie Goes to Monte Carlo* (a city that had a Chevrolet named after it). It was a game of diminishing returns: *Herbie Goes Bananas* of 1980 is filled with "south of the border" clichés and stereotypes.

Disney produced what may be the ultimately apt Herbie: a plush toy version of the car, without structure or skeleton. Herbie the Love Bug made his way into Disney's *World on Ice* show, where he was given a mouth under his hood, complete with teeth and tongue. A television series flopped after only five episodes in 1982, but by 1997 there was a made-for-TV remake, and a feature on the way for 2002.

Disney's imagineers later supplied a Herbie theme for a hotel at Disney's All-Star Movies Resort in Florida. A huge Herbie stood beside a giant upright tire and sculptures of giant tools, three-story-high wrenches and pliers cribbed from Claes Oldenburg. The hotel wing was decorated with a frieze of checkered flags.

Herbie was a character so rudimentary that it had less personality than most Disney cartoon figures. Other movie cars had real character, like the animated Bennie the Taxi in *Who Framed Roger Rabbit* of 1988. But the best film cars traveled into darker territory. John Carpenter's *Christine* was based on Stephen King's novel about a teenage boy's love affair with a demanding and jealous 1958 Plymouth Fury. The model was important: 1958 marked the pinnacle of the tail fin era. Chrysler, Plymouth, and Dodge offered the most extreme fins with its "Forward Look," created by Virgil Exner, who, according to some stories, had designed the Bug's sibling, the Karmann Ghia convertible, while he was between jobs at Pontiac and Chrysler.

To be possessed by a car—that was a truly American theme. But in the most telling moment of the film, Christine's primal injury came from a worker at the assembly line. It had become urban legend that workers angry at conditions in the factory would insert noise makers or create intentional rattles. But Christine retaliated against a worker; unlike Herbie, she was a car in control.

In many ways the most revealing of the Herbie films is the 1997 television version. Here, at last, is a metaphysical parable. Herbie is a living thing because he is suffused with love in the form of the picture of the engineer's beloved wife that falls into the molten metal from which the car is built. There's a backstory of Herbie's origins: Herbie's creator, a benevolent German scientist, was picked up by the Americans after the war, who misunderstood what he had done (a wry parable here): "They thought my little people's car *was* a people."

The scientist is forced to produce a second living Beetle, but it becomes evil—*"Ungeziefer"* he says, German for vermin. The anti-Herbie is named Horace. With vents on his front hood, of sinister effect but mysterious function since the engine is still in back, this Hate Bug is driven by his creator's self-love. Horace the black Beetle looks a lot like the first KdF prototypes, right down to the blackout shades on the headlights.

The Hate Bug destroys the Love Bug: Herbie is about to be buried in a huge crate with one hubcap on top (one of the few times in the Herbie cycle where one actually glimpses the VW logo). But a last-minute plea intervenes: he is rebuilt.

The film's climax as always is a race—Hate Bug versus Love Bug this time. Horace turns out to be packed with secrets, James Bond–style weapons—a drill that emerges from inside its axle (a patent theft from *Ben-Hur*'s chariot race), a grenade launcher, and a powerful laser that cuts metal. But even sliced in half, Herbie wins. The evil Horace careens off a cliff and in a moment worthy of Nathaniel Hawthorne, if Hawthorne had written for Disney, crashes into the ground and, to judge from the flames that blossom up, falls straight to hell.

The real Bad Bug, the Evil Beetle, the Hate Bug took other forms. During those same days in March 1969 when America's children were laughing at Herbie's antics, Charles Manson, the 1960s' most notorious psychopath and most potent symbol of the dark side of the counterculture, also fell in love with the Bug.

Peeling off $1,300 from a thick roll of hundred-dollar bills, he bought his first dune buggy from Butler Buggies on Topanga Boulevard in Los Angeles. Before long, the Manson "family" was stealing other buggies and building them.

The group of runaways and dropouts clustered around several ranches whose owners their leader had managed to charm or intimidate. One was a frequently used film location known as the Spahn Movie Ranch. Spreading old parachute canopies from trees as shelter and using a stolen generator, the Manson gang created a miniature assembly line for converting Volkswagens into dune buggies.

Far more remote was the Barker Ranch near Death Valley. Located up a narrow canyon filled with boulders called Goler Wash, it was only accessible by four-wheel drive and figured in Manson's fevered imagination as his ultimate redoubt—a kind of Führerbunker for his Götterdämmerung. (Manson once drove a school bus up the narrow canyon but was never able to get it out again.)

Manson and family hung around with a motorcycle gang called the Straight Satans, and he considered mounting the family on cycles but decided the buggies were more practical in the desert. With special oversized gas tanks their range exceeded a thousand miles.

Manson was possessed by visions of rock and roll stardom and apocalyptic leadership. He latched on to Dennis Wilson of the Beach Boys, whose music was wrapped in the California car culture of surf and sun, sports cars and woodies and dune buggies. Wilson tried to get Manson a recording contract.*

---

* The Beach Boys actually recorded one of Manson's songs, "Never Learn Not to Love," which previously bore Manson's less upbeat title "Cease to Exist," on their album, 20/20, released in February 1969. The song begins "Cease to resist, come on say you love me/ Give up your world, come on and be with me." Manson was not credited in the album liner notes.

When that failed, Manson's visions turned darker. He imagined a California apocalypse launched by a violent African-American uprising. He called it Helter Skelter, a phrase borrowed from the Beatles song of the same name. For Manson, Helter Skelter was a welcome vision.

As part of this apocalypse, Manson said, he imagined himself as a kind of Rommel of the American desert. His vision was also colored by the Book of Revelation. The dune buggies were to be the "horses of the apocalypse" with "breastplates of fire."

Manson's own dune buggy was outfitted as a command wagon, with a holster for his pistol and a scabbard for his two-foot sword. It had ocelot fur upholstery and was painted in an informal camouflage created by scattering sand on wet brown paint.

Manson's dune buggy obsession is described in a very strange book based on extensive interviews with family members and local witnesses. Contradictory and confused, *The Family: The story of Charles Manson's Dune Buggy Attack Battalion* by Ed Sanders, a former member of the Fugs rock group, nonetheless captures the delirium of Manson's fascination:

> From his experience in the rough terrain of Death Valley, Manson decided that dune buggies were just the vehicles for his mobile snuff squad. They were great for outrunning cops in the abyss. They were light enough so that two or three of the gore groupies could lift them over boulders and precipices. Motorcycles, on the other hand, were scorned as being inadequate in the wilderness. But dune buggies, ah sacred dune buggies—they were like battleships. He would later outfit dune buggies with huge gas tanks giving them a 1000-mile assault field. They put machine-gun mounts on them and Manson's command buggy was fixed so that it could be slept in. There could be food dune buggies, ammunition dune buggies, dope supply dune buggies, etc.

On July 15, 1969, a Los Angeles sheriff's officer flew over the desert north of the city in a helicopter and saw three Volkswagen floor pans lying near the Spahn Movie Ranch. He suspected car thieves but did nothing.

Manson was afraid the police were watching for his dune buggy, so the gang drove in a yellow 1959 Ford to the Tate and LaBianca homes in August 1969 when they committed the horrifying murders that symbolize the 1960s gone wrong.

The dune buggies themselves have been kept alive in bits of Manson lore like rock star Marilyn Manson's 1991 song "Dune Buggy."

Herbie and Charlie—the pairing represented a mad polarization of the Bug's basic gestalt. Herbie was the car's soft shell, as cute and innocent as a plush child's toy. Charles Manson's death buggy was the hard skeleton underneath, the very embodiment of evil in a vehicle.

The very real usefulness of the dune buggy would be extended by the Chenowth company, whose specialized dune buggies attracted the attention of the Pentagon. Like agile and angry Jeeps, they served well for spies and commandos. The army began to equip special operations and reconnaissance teams with the dune buggy assault and reconnaissance vehicles. The first vehicle to reenter Kuwait City during the Persian Gulf War of 1991, the company proudly boasted in its promotional literature, was a Chenowth dune buggy:

> The first U.S. forces . . . rolled in on the same high-performance machines that win prestigious off-road races such as the Baja 1000.
>
> Under eerie skies darkened by smoke from burning oil fields, the desert racers, painted black and equipped with guns, climbed over roadblocks and scaled 8-foot-high berms to penetrate the ravaged emirate.
>
> Because the hastily constructed obstacles could not hold back the off-road vehicles, the SEALs were able to go where they wanted to and avoid traps left by fleeing Iraqi troops. One Kuwaiti citizen dubbed the vehicles the "Ninja cars."
>
> The vehicles, called Fast Attack Vehicles, or FAV's, raced ahead of U.S. troops to scout out territory, and darted behind enemy lines to assess the size and position of enemy forces.

The company produced an improved version, the Advanced Light Strike Vehicle, which a brochure bragged "features a main weapon station, with 360° arc of fire, designed to host the M2 .50-caliber machine gun or the MK19 automatic grenade launcher. The ALSV can also utilize remote control and stabilized platforms to provide accurate shoot-on-the-move lethality."

The Chenowth Advanced Light Strike and Fast Attack Vehicles were the literalization of *Kustom Kar Kommandos,* part weapon, part cartoon car. Individuals could buy a Chenowth dune buggy pretty much identical to the FAV, sans weapons. And Mattel gave its G.I. Joe soldier toy a version of the FAV to supplement his jeep and Hummer.

The successor of the Kübelwagen was used to keep Kuwait free and guarantee that Americans could count on at least another decade of dollar-a-gallon gasoline.

In 1970, readers of the counterculture bible, *The Whole Earth Catalog,* browsing through the write-ups of ecotoilets, fruit-drying devices, and geodesic dome kits would have come across the announcement of a new book: *How to Keep Your Volkswagen Alive.* Full of chatty talk about how to change a spark plug or adjust a timing chain, accompanied by illustrations straight out of the underground press, the book soon became a best-seller—and went on the hippie top ten list.

The book's author was John Muir, a long-haired mechanic with a garage in Taos, New Mexico, and a descendant of John Muir the naturalist. In folksy explanations, with cartoonlike diagrams, Muir summarized the advice he had been giving to his customers.

Muir had worked as an aerospace engineer—one of his last jobs was with Lockheed missiles in Sunnyvale, California—before dropping out and moving to New Mexico where he became part of the local hippie scene. His wedding in the summer of 1968 was attended by 500 people, including Wavy Gravy and the entire Hog Farm commune.

Tired of dealing with people who sought his advice when the garage was closed, Muir decided to publish a book of his wisdom about fixing

Beetles. During the summer of 1969, while men landed on the Moon and Teddy Kennedy landed in the water off Martha's Vineyard, and while hundreds of microbuses headed for Woodstock, replicating the Ken Kesey Magic Bus experience, Muir and his wife, Eve, stayed home to finish the book.

Muir's key mechanical advice was don't lug the engine, that is, avoid the American tendency to shift too quickly to higher gear. Let the engine work where it is happiest, in a lower gear, where it will pull steadily and go on forever. He understood that Porsche and Reimspiess had designed the motor to be run right around 2,000 rpm almost forever, and that pushing the engine too much would damage it.

The lesson was a metaphor for the hippie life. There were others: treat the car like a donkey, a horse, he said. And do all the work on the car *with love*.

In Muir's book, the Bug came alive in a different way than in the Herbie movies. Like any experienced engineer, he knew that mechanisms were full of aberrations in behavior, gremlins and glitches—bugs.

The concept of the living bug in the book was that of a hippie philosopher. Unlike Robert Pirsig, whose *Zen and the Art of Motorcycle Maintenance* faded from pop Zen into misunderstood Plato, Muir's advice was practical as well as philosophical. Taking it easy meant not overtorquing bolts—American power tools tended to tighten fasteners on a Bug too much. Don't overtorque also meant don't get wrapped up too tight. Low torque, low gear—that was the mode of life for an engineer who dropped out of the rat race, keeping his own revs down.

The book's content was a combination of folk and Zen proverbs. The book's epigraph was "Come to kindly terms with your ass for it bears you." It was full of hippie wisdom. "Avoid shortcuts the same with long cuts. . . . Talk to the car, then shut up and listen. . . . Feel with your car. . . . While the levels of logic of the human entity are many and varied, your car operates on one simple level and it's up to you to understand its trip. . . . Use all of your receptive senses . . . the type of life your car contains differs from yours by time scale, logic level, and conceptual anomalies but is 'Life' nonetheless. Its karma depends on your desire to keep it—Alive!"

"The real enemy is Rust," Muir wrote in one passage. "Every morning a dew settles over the earth urging plants to meet the sun. This same dew settles on your precious VW, coating it with moisture." The result was rust and the antidote Muir wrote was love. "Love keeps alive . . . love of things mechanical . . . love of things the way they should be."

To illustrate the book Muir found another hippie not far from Santa Fe. Peter Aschwanden thought of himself mainly as a painter who did drawings and illustrations to support his art. He was also a fan of cartoonist R. Crumb. "That big foot thing made quite an impact," he said, referring to the famed "Keep on Truckin'" image that was Crumb's best-known work.

Aschwanden was born in East Los Angeles. His mother, a refugee from Europe, had seen a prewar Bug and was startled when the same car showed up in Los Angeles. To Aschwanden, the Beetle was an urban phenomenon like the counterculture and the world of the hippie. Aschwanden had tried driving a Volkswagen bus in the hills around Santa Fe but soon got tired of straining uphill at 30 miles an hour and bought a succession of Ford pickups.

Aschwanden's drawings of the upright, straightforward VW engine and other mechanisms were so lightly and happily done they could hardly intimidate the user of the book. In one all the parts of the car were laid out as if for sale at a bazaar. Another image, later reproduced on a poster many Bug owners treasured, was a bird's-eye view of the car with passengers in place and the shell of the body cut away.

Aschwanden recalled the power of Muir's ideas and the communal spirit that produced the book. "There was a lot of energy," he said. "It was coming together. The time was in sync."

Muir wrote a sequel of sorts, *The Velvet Monkey Wrench,* which tried to apply the greater wisdom Muir had learned fixing Beetles to reorganize the polity. It was *Zen and the Art of Motorcycle Maintenance* meets Plato's *Republic.* After his death, Muir's wife, Eve, offered a philosophy in the updated edition of the book that extended the ideas he developed to nothing less than human nature.

"Each of us," she wrote, "has inside us a jewel and sometimes it gets muddy and its facets are dulled by repetition." It was a jewel on a gimbal, like a jewel inside a watch, she explained, and for things to be on time, in sync with time, the jewel has to be clean. "Sometimes people get out of sync with time. Sometimes the jewel gets gummed up. Then you just put a little WD-40 lubricant on it."

Fixing a life, fixing society—it all could be as simple as fixing a Bug.

chapter eleven

n his 1973 time travel comedy *Sleeper,* Woody Allen comes across a dusty, neglected Beetle. He hops in, turns the key—and the car starts right up, after centuries of neglect. It could have been one of VW's own ads. But the scene hinted at another truth: that by the early 1970s the Beetle had become something for a time capsule. It lived on, freeze-dried, in a kind of suspended animation, and while on February 2, 1972, the 15,007,034th Volkswagen Beetle was manufactured, making it the most successful car of all time, the car was selling largely on the strength of its myth. But not for long.

In West Germany sales had been declining for more than a decade. On the morning of July 11, 1974, the last Beetle to be produced in Wolfs-burg rolled off the line. Production continued at the factory in Emden, West Germany, until January 1978. Beetles were still built in Puebla, Mexico, and São Bernardo do Campo in Brazil and at a new plant in Lagos, Nigeria, which went on line in 1973. The last year the Beetle sedan was sold in the U.S. was 1977; the convertible remained popular a while longer—the fun model significantly outlasting the practical one.

But the Beetle was firmly bound not just into West German culture but into cultures all over the world. It acquired nicknames everywhere: Käfer in Germany, Kever in Holland, Coccinelle in France, Maggiolino in

Italy, Escarabajo or Pulguita ("Little Flea") in Spain. But Poles ("Garbus") and Filipinos ("Kotseng Kuba") associated the car with a "hunchback." In Finland it was a bubble or Kupla. The British used Beetle more than Bug; Americans were split about evenly.

The idea of the Bug separated from its reality as practical product and existed as an idealized collective memory. In the 1960s, Pop artist Tom Wesselman had rendered Beetles in big juicy colorful paintings. Andy Warhol silkscreened the famous "Lemon" ad—in effect making an image of a magazine page of a photo of a real car—revealing it as part cliché, part icon, as he had done with Jackie Kennedy and Marilyn Monroe, car wrecks, electric chairs, and other subjects.

The popular car became Pop Art but pop cartoon as well in toys and collectible knickknacks and such designs as the Volkschair, an easy chair shaped like a bug, created by a witty Yale grad student, Doug Michels. Michels later joined the artists collective Antfarm, which in 1974 constructed the Cadillac Ranch sculpture near Amarillo, Texas, a monument to the energy crisis and the demise of the tail fin ideal.

By the mid-1970s, the Beetle appeared in the deadpan photorealist paintings of Don Eddy, in parking lots or curbside. The colors were muted, even wan, and the paintings showed an overly sensitive awareness of the reflections on the car bodies and a certain lethargy of spirit: Eddy's was a too bright, morning-after vision—part of the great hangover from the 1960s, modernist abstraction, and Pop Art.

As Pop gave way to a new realism, the counterculture was absorbed and turned into a checkout counterculture, its ideas and images absorbed and played back by ad men and marketing mavens.

VW attempted to tap into enthusiasm for the dune buggy in 1973 by offering the vehicle it called the Thing. The name evoked "do your own thing" or Thing, the hand-shaped creature in the television show *The Addams Family*. The letters on the side were psychedelic, like those on a *Love Bug* poster, vaguely mushroom-shaped, but the body and the car were essentially a replication of the Kübelwagen. The VW Thing was an anti-Bug, and no one with any hint of irony could have produced it.

In 1973, too, came the publication of a book called *Small Is Beautiful*. Its author, E. F. Schumacher, was a mild-mannered German-born British

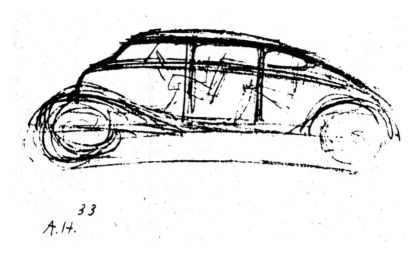

33
A. H.

Hitler offered detailed advice in the design of the Volkswagen, including sketches.
(all photos courtesy of Volkswagen AG, except where noted)

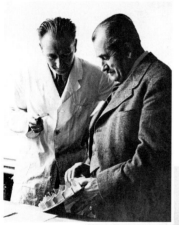

LEFT: Ferdinand Porsche (right) presided over the creation of the Volkswagen, while Franz Reimspiess designed both its engine and logo.

BELOW: Volkswagen prototype, 1936

In May 1938, Hitler presided over the ceremonial laying of the cornerstone at the vast VW factory.

To promote Volkswagen sales, prototypes were paraded through Berlin in 1939.

ABOVE: Porsche's engineers created a scale model of the VW to give to Hitler on his birthday in 1938. Porsche (left) and Göring (far right) look on.

LEFT: Hitler hoped the people's car would be as successful as the people's radio, or Volksempfänger. (*photo courtesy of the author*)

Der KdF Wagen

Promotional images for the KdF-Wagen showed a happy family on the autobahn.

The air-cooled Kübelwagen, a militarized VW, proved as rugged in the cold of the Eastern Front as in the heat of North Africa, where it won Rommel's praise.

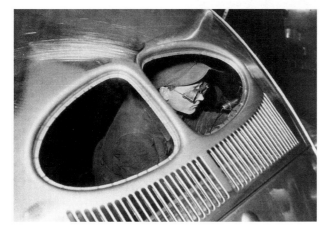

Within weeks of the end of the war, British engineers resumed production of the Beetle on the assembly line at the Wolfsburg factory.

Doyle Dane Bernbach's "Think Small" ad became the most famous in American history.

Think small.

Our little car isn't so much of a novelty any more.
A couple of dozen college kids don't try to squeeze inside it.
The guy at the gas station doesn't ask where the gas goes.
Nobody even stares at our shape.
In fact, some people who drive our little

flivver don't even think that about 27 miles to the gallon is going any great guns.
Or using five pints of oil instead of five quarts.
Or never needing anti-freeze.
Or racking up about 40,000 miles on a set of tires.
That's because once you get used to

some of our economies, you don't even think about them any more.
Except when you squeeze into a small parking spot. Or renew your small insurance. Or pay a small repair bill. Or trade in your old VW for a new one.
Think it over.

The bus, the Bug's sibling, served as the unofficial vehicle of the 1960s counterculture. VW ran this ad after the death of the Grateful Dead's singer and guitarist Jerry Garcia.

Jerry Garcia. 1942-1995.

The Bug's cousins: the Fiat Topolino, or Little Mouse, arrived in 1936. *(photo courtesy of Fiat)*

VW offered the beach buggy version called the Thing in 1973. *(photo courtesy of the author)*

For many, the new people's car, circa 1999, was the Daimler-Chrysler Swatchmobile, or Smart car. *(photo courtesy of Smart/ DaimlerChrysler)*

The Beetle's shape was reborn in 1993 in the Concept One, seen in front of VW's Simi Valley studio.

Concept One designers J Mays (left) and Freeman Thomas

The New Beetle was introduced with great fanfare in Atlanta (with the de rigueur Coca-Cola signs) in February 1998.

Underneath the New Beetle's geometric body lay the same basic car
as the Golf, Jetta, and Audi TT.

The European RSI
racer version showed
that the New Beetle
could take on a
tough rather than
cute personality.

In 2000 VW opened Autostadt, the theme park/museum built on the site of wartime barracks beside the factory.

Ferdinand Piëch, grandson of Ferdinand Porsche, became head of VW in 1993 and approved the production of the New Beetle.

At the factory in Puebla, Mexico, workers assemble both new and old Beetles beneath religious shrines. *(photo courtesy of the author)*

economist who worked for the British Coal Board and wrote essays on "Buddhist economics" in his spare time. He had helped rehabilitate Germany after the war and served in the British Control Commission at the same time the British officers Hirst and McEvoy were rehabilitating the VW plant in Wolfsburg.

The vaguer sentiments of the counterculture had by the 1970s taken on the systematic form of policy proposals: emphasis on smaller scale, reduced energy demand, renewable industrial processes, and more durable products. Schumacher's book provided intellectual rationales for many of these ideas, championing the benefits of small-scale enterprise over massive enterprise and local innovation over mass production. "Economics as if people mattered" was the subtitle. He argued that traditional economics had focused too much on hard numbers such as the GNP and such unexamined values as maximizing consumption. In that pursuit, other choices and more humane values had been forgotten. He and others urged a shift in technology from mega-projects to "human scale" values. No more dams, but more village schools and electrical generators.

The energy crises of the 1970s forced mainstream politicians and economists to consider issues previously located on the countercultural fringe: dependence on non-renewable fossil fuels, fuel efficiency, and car size. Schumacher was embraced by President Jimmy Carter, and Governor Jerry Brown of California established an Office of Appropriate Technology. His ideas were echoed in a number of books arguing for simple technologies, such as Kirkpatrick Sale's *Human Scale* and designer Victor Papanek's *Design for the Real World*. Usually, appropriate for these writers meant simpler, cheaper technology more accessible to the average person or to Third World economies. "Appropriate technology" might have applied at one time to torsion bar suspension and air-cooled engines, those simpler and cheaper alternatives to the dominant automotive technologies of 1938, and, they argued, it remained a viable option in 1978.

The Beetle's technology continued to provide practical transportation and a democratic symbol in the Third World. Emperor Haile Selassie of Ethiopia claimed to have introduced the first automobiles to Ethiopia in the 1920s and for decades most of them belonged to him. The King of Kings, Elect of God, Lion of Judah, and Most Puissant Majesty owned a fleet of Rolls-Royces, Mercedeses, and Lincoln Continentals. But as Ryszard Kapuscinski reported in *The Emperor,* it was not without symbolism when the Ethiopian army overthrew Selassie on September 12, 1974, that a Beetle appeared at the palace to carry him away.

"You can't be serious!" the Emperor bristled. "I'm supposed to go like this?" An army officer folded the front seat forward and finally, reluctantly, Selassie got into the back seat and was driven away into exile.

In the United States, beautiful was still large. Since World War II, Detroit has rarely put its heart, much less its best technology, into small cars. The second generation of small cars they marketed to compete with the Beetle were worse than the first. The names Pinto and Vega live on as symbols of Detroit's incompetence. First planned in 1968, the year of the Beetle's highest sales, the Chevrolet Vega arrived on the market in 1971 nearly 400 pounds heavier than the Beetle and $300 more expensive. The car was manufactured at the infamous Lordstown, Ohio, plant where GM attempted to install its first robot assembly line, prompting a strike in 1972. Workers angrily hid rattlemakers—old bottles and cans filled with nuts and bolts—in the innards of the vehicles.

That the Beetle still sold was testimony more to the failings of the American competition than to its own virtues. Volkswagen desperately needed a successor to the Bug. Volkswagen's bosses clung to it for too long, as Ford had clung to the Model T, an idea they could not shake.

Throughout the 1950s and the 1960s VW management developed many variants on the Beetle with different bodies, such as the Type 3 and the Fastback, but they could never quite bring themselves to replace it.

VW built a vast test center at Ehren-Lessien, north of Wolfsburg, with steep hills and highly banked turns, cobbled stretches, loops and straightaways, and the most sophisticated wind tunnel in Europe, like a great jet engine. But few of the cars tested in these state-of-the-art facilities came to market.

In their paralysis, VW executives arrogantly ignored the doubts and suggestions of others, especially Americans. They had remained deaf to the cries of U.S. dealers for nearly fifteen years and refused to bring in automatic transmissions and larger engines. Air-conditioning and radios lagged. They pushed such models as the Dasher/Quantum and the Fox on American dealers, but they sold poorly. This arrogance was made manifest in the new VW factory that opened in Westmoreland, Pennsylvania, in 1978. The manuals were written in German and the machines and fasteners were metric. Within weeks there was a strike.

The executives also could not understand why the Jetta notchback sold so well in the U.S. and was preferred to the Golf hatchback, which was dominant in other markets. But hatchbacks were associated in the U.S. with the cheap low-power econoboxes, like the shoddy Chevrolet Vega and Chevette that energy crises had inflicted on American buyers.

When at last the Beetles' successor arrived—the Golf, or, as it was called in the U.S., the Rabbit—it was a more conventional car. And of course it faced a hopeless task. What could follow such an act as the Beetle?

Introduced in the U.S. in 1975, the Rabbit was more like contemporary Japanese cars than like the Beetle—the Hondas, Toyotas, and Datsuns (later Nissans) that had moved ahead of Volkswagen in U.S. sales in the early 1970s. It had front-wheel drive and a water-cooled engine. (With the U.K.'s Mini in 1959, the water-cooled transverse-mounted motor had become the standard configuration for small cars around the world.)

Visually the Rabbit was the direct opposite of the rounded Beetle. The famous Italian designer Giorgio Giugiaro, known for his Maseratis and Ferraris, had created a car that seemed to have been cut out and assembled from flat pieces, like a cardboard model of a car. Its design was described as "origami style."

It was a good enough car, and sold well in Europe as the Golf, but it suffered from quality problems. It was no Beetle.

Also in the 1970s, the Bug had to answer a devastating challenge to the safety of the car from Ralph Nader and his colleagues at the Institute for

Automobile Safety. Nader began his crusade in 1959 with an article in *The Nation* attacking auto safety. He rocked the automobile industry in the mid-1960s with the book *Unsafe at Any Speed: The Designed-in Dangers of the American Automobile*. Its first sentence became famous: "For over half a century the automobile has brought death, injury and the most inestimable sorrow and deprivation to millions of people."

Revealingly, Nader spent most of his energy attacking not the traditional Detroit car with its tail fins and bumper bullets, its huge motor and weak brakes, but the Chevrolet Corvair, the best and most innovative of the Detroit compacts. The Corvair had been inspired by features of the Bug: rear engine, air cooling, and independent rear suspension. It was this suspension that made the first Corvairs unsafe, but by the time Nader's book appeared, the problem was essentially fixed.

After General Motors assigned private detectives to find compromising facts about his private life, of which they found very little, Nader won a suit for damages and used the money to fund additional research into automotive safety.

In 1972 Nader's next book, *Small on Safety,* laid out in exacting detail the safety flaws in the Volkswagen. Engineers and drivers had been aware of the car's defects since the 1930s, of course: its vulnerability to cross winds, its oversteer in curves, and its thin shell construction that made it vulnerable to the impact of large American cars and trucks. But Nader's researchers had worse news for Bug drivers.

The book was a precise document, like the testimony of an engineer serving as expert witness at a damage trial. It came complete with diagrams of the VW fuel filler cap, which would allegedly belch gasoline in an accident, ready to ignite with a spark from any shorted wire. Dotted lines marked the likely path of ejection through the rear window of a driver whose seat came loose. Diagrams of the door locks and catches, which the text condemned, were precise enough to satisfy the patent office.

Nader did not attack the overpowered and underbraked Cadillac or Lincoln, but focused on models likely to be owned by influential intellectuals, his intended audience. The message was implicit: if the Beetle is this dangerous, how much worse is the average car? Targeting the innovative Corvair and the Beetle suggested Nader's real target was the whole national obsession with the automobile. He was in the forefront of a growing discontent with the automobile that would lead to regulations establishing safety standards and emission levels, laws that would eventually result in the demise of the Beetle in America.

Nader's attack helped crystallize an emerging view of the automobile not as dream machine, but as transportation appliance. A significant number of Americans now viewed their cars not as an exciting means of self-expression but as a mere necessity, even a necessary evil. In a sense Nader completed a debunking process VW had begun in its ads. After he came on the scene, a whole structure of government and nonprofit agencies, publications, and consumer advocates grew up that tested and rated cars like television sets or toasters, charting durability, quality, and value. This approach was perfectly tailored for the dependable but unexciting cars the Japanese were beginning to export to the U.S.

The Japanese cars offered the Beetle serious competition. They were not just more modern than the Bug, with equal or greater economy and durability, but were manufactured in a new, more efficient way.

In 1970, when VW reached a highwater mark in U.S. sales of more than 569,000, Toyota was selling about 125,000 cars in the United States—about the same number VW sold in 1958. A year later, Toyota sales rose to 200,000 cars.

Japanese carmakers challenged the basic methods of River Rouge and Wolfsburg. Their theory of manufacturing turned on the belief that efficiency and quality were not opposite, but complementary, goals. The Japanese worker on the line was also the quality inspector. Instead of pushing the assembly line to top speed, fixing defects at the end, the line was stopped until problems were solved. The Japanese system constituted an implicit rebuttal of the famous VW "Lemon" advertisement, which showed an apparently perfect Bug rejected for a blemish on the chrome strip on the glove compartment.

Other VW ads had boasted of the number of cars that were rejected and of the 3,389 inspectors at the Wolfsburg factory—more than one for each of the 3,000 cars produced there each day. Such claims sounded good in 1960, but by the end of the century, the Japanese system, which became accepted wisdom in worldwide automobile production, held that quality control should be integrated with manufacturing. The number of quality inspectors in a factory was no longer something to brag about.

James Womack and a team of researchers from MIT undertook a famous investigation of the auto industry in the 1980s. Their landmark study was called *The Machine That Changed the World*. They found that the West Germans were spending as much time in post-production inspections and fixes as the Japanese spent to build the cars in the first place. Their findings challenged the value of the craftsmanship of old-style manufacturing, the image of European workers somewhere between old-world watchmakers and Black Forest elves. Kurt Kroner, the inspector mentioned by name in VW ads, was obsolete.

The Japanese system was based in a business culture that, like West Germany's, had been reshaped by Allied occupiers.

To thwart the rise of leftist influences in Japan after World War II, the Americans encouraged an agreement between management and labor unions. The unions traded their right to strike for a guarantee of lifetime employment. The result was that plant managers could no longer treat labor as a variable—hiring and firing with demand—as in the traditional mass production system. Labor became a fixed cost and managers were given an incentive to make it as productive as possible. Workers became more: they became the foremen and inspectors as well. The system produced other economies: in the so-called lean, or just-in-time, system, parts were brought to the line at the last minute, reducing inventory costs. And quality rather than quantity became the first priority of the assembly line. The Japanese companies were also able to do this because they were not tied to quarterly profit reports, like American companies with their many stockholders. They sought market share above profits; Toyota and Nissan laid out plans over decades, not months or years.

In West Germany, too, agreements between management and labor had prevented the boom-and-bust hirings and layoffs characteristic of the American system. But no real partnership existed, only the shared benefits of producing more and more Bugs each year. When sales began to level off, the harmony vanished.

The 1970s brought an end to VW's proud and special status as a national resource. The belief of the postwar managers that the company would "belong to the German people" turned out to be dreamy idealism.

Mitbestimmung, the cooperation of management and labor, with each having a voice in the direction of the enterprise, had formed the foundation of the economic miracle. It worked so long as growth continued. For a long time, the menace of the East, the desire to rebuild the country, and the personal power of the paternal Heinz Nordhoff made it work. Similar arrangements governed most of German industry. But *Mitbestimmung* contained the seeds of the later problems.

As early as 1954 the first strike had broken the peace, when autoworkers in Bavaria had dared to challenge the truce between labor and management. After VW became a joint stock corporation at the end of the 1950s, it remained partially owned by the federal government and the state of Lower Saxony. Seats on the board went to labor leaders—notably the powerful leaders of the works council—and government leaders for whom job cuts were politically intolerable. Labor and management in effect held equal power in the direction of the company.

By the 1970s, which brought Japanese competition, a series of energy crises, and rising wages, *Mitbestimmung* began to paralyze VW and other German companies. By the 1970s, Wolfsburg's efficiency was below that of Japanese and even American factories. Contracts tied to inflation increased wages without a corresponding rise in productivity. Politics prevented layoffs or changes in work practice. Part of the problem lay in management's lack of understanding of the Japanese challenge, part in the power of the German unions, especially I.G. Metall. This national industrial workers union, often called the most powerful union in the world, was closely linked to the Social Democratic Party, which held national power from 1969 on.

Management and labor found themselves locked, exhausted, as in a boxer's clinch—a form of industrial funk that became epidemic in the 1960s at British Leyland, Fiat, and Renault as well as at VW.

In the 1970s, automakers sought to curb labor costs by trading off job security for wage freezes. With less flexibility to lay off workers, Europe and the U.S. naturally looked to the Japanese system, which had been designed to deal with that situation. But first Western managers had to abandon mythologies of "high labor costs." For years, they had resisted adopting innovations, blaming the success of Japanese cars simply on "cheap labor." In fact, the Japanese system conserved labor. Japanese firms required roughly half as many hours to produce a car as Volkswagen and other European factories did and a third as many hours as U.S. plants.

Western managers failed to understand that the key to the Japanese, or lean, system lay in just-in-time delivery of parts, to save warehousing costs, as well as in labor methods. Before adopting the Japanese system, managers first sought a solution in automation, resorting to robots they called "iron slaves." But in Wolfsburg, in the much touted Hall 54 where the robots were installed in 1983, the iron slaves were constantly making errors, and the Kurt Kroners of the factory had to clean up after them.

The introduction of robots and other automation technology naturally bred resentment. The secret of Japanese success, it soon became clear, the answer to the labor problem in the Western Fordist factory, was not more machinery but more humanity.

Trust the worker, provide parts just in time to cut inventory costs, and both quality and quantity of production would increase. Soon in automobile factories all over the world the adoption of the Japanese system was clearly evident in its symbol—the rolling stacks of the brightly colored kanban, or parts bins, standing beside the workers.

Outside the factory, too, cultural changes affected the Bug. Small is beautiful took on new meanings in the 1970s, as a technological ethos developed that valued miniaturization. In electronic products, small size, portability, and personality were interlinked. Things that could be carried around tended to be more dramatically designed. In his book *Populuxe*, Thomas Hine noted how when most American families acquired a second television set it was a portable. In the late 1960s Sony offered what its advertisements termed "the tummy Trinitron," a color set small enough to sit on a person's stomach.

With the advent of the transistor, the people's radio turned into the personal radio and the portable radio. Although the transistor was developed in the United States and the first transistor radios were American, companies like Sony made them the quintessential symbol of Japan in the 1950s and 1960s.

Sony's TR-63 radio, introduced in the U.S. in 1957, was advertised as small enough to fit into the pocket of a man's shirt—and when the engineers didn't quite achieve that goal, Sony ordered up shirts specially tailored with wider than average pockets for its salesmen.

There was a fitting symmetry to the fact that 1978, the year the Walkman came on the market, was the same year that VW's German factories produced their last Beetle. The Walkman was another form of universal appliance: a successor to the Brownie, a Model T for the ear. The people's radio was succeeded by the personal tape player, which, in turn, was succeeded by the personal computer and personal digital assistant. "Personal" was a salesman's buzzword, but the designers made it good by giving the devices visual personality. The Walkman's very name suggests its anthropomorphization and Sony's billboards depicted the tape player as a dancing little black slab with arms and legs.

Small was beautiful—and Japanese. The Honda Civic compressed high technology into a small package just as the Walkman did. Higher technology meant more tightly integrated circuits as electronics took on such tasks as regulating engine efficiency. Tiny radios, tiny cell phones, tiny laptops, and pocket-sized digital assistants would take on the prestige value once held by longer cars and bigger television screens.

In the mid-1970s, when Steve Wozniak sold his VW van to raise a few hundred dollars to set up Apple Computer with his friend Steve Jobs, it marked a turning point—from counterculture to computer culture, from microbus to microcomputer.

Apple's Macintosh aimed to make a computer as friendly as a Bug. "Hello," the screen opening read, "I'm Macintosh." Steven Levy in his book on the Macintosh, *Insanely Great,* explains that the single-box, self-contained Mac was designed to be as universal an appliance and as basic a shape as the Beetle. It was to be a Volkscomputer, "the computer for the rest of us," but by the mid-1980s, the real Volkscomputer turned out to be the IBM PC clone—cheap, standard, accessible—and made mostly in Asia.

chapter twelve

by the 1980s the Beetle had become a piece of the universal cultural code, a kind of gestalt. And for VW it remained a ghost that haunted the company. Even before the last Beetle was sold in the U.S. in 1979, owners had established a Vintage Volkswagen Club. And the company itself seemed to view the Beetle's amazing success from a distance: "Our car, the movie star" a 1974 ad marveled, showing a Beetle wearing huge sunglasses.

Part of the subtext of the 1960s ad campaign had been that the Bug was a product beyond aesthetics. It was the quintessential modernist argument: design based on function goes beyond style. Just as the improvements were invisible things, the decisions that went into the design were those of engineers, not stylists—or so we were to believe.

And if it was simply a piece of engineering, as the legend held, it was also an empty vessel, ready to accept individual projections and fantasies. No wonder in the 1980s the Bug became a favorite of "art car" buffs, who planted grass on it, painted it, or rendered it in wood.

The car's life in the cultural consciousness was grounded in specific personal images and memories. Its imaginative existence was laid out in millions of small experiences—sights framed by the pre-1953 split or pretzel window, the characteristic sound of the engine, ice cream spilled on seat fabric, a kiss stolen across the gear shift. Or cold hands thanks to the weak heater and a fragment of highway seen through a hole in a rusty floor pan. The owner's affection was often based as much on the car's deficiencies as its efficiencies, just as we often love a person more for his vices than for his virtues. From the Model T to the Bug and beyond, failings could be humanizing.

Out of such experience grew the collective imagination of the Bug. Tales about the Bug were versions of folklore—passed along by telling and retelling. "Everyone has a story" is the familiar line. "Everyone owned one or had a brother or cousin or friend or neighbor who owned one."

The Beetle had become a "type form," as designers call a fixed and settled design, for the automobile—a near archetype. This was true even though very few automobiles literally looked like the Bug. The tip-off was in the shape of children's toys, illustrations for children's books, and, indeed, pictures children drew.

Just as a child will draw a house with an angled roof and smoke puffs coming from the chimney even if he or she lives in a flat-roofed modern house with no fireplace, the cars children sketch tend to be round, with smiling faces, like the Beetle. Consider the similarity of the Beetle's face to that of Putt-Putt, the car in computer games, or Bumpy, the car in children's books.

The Beetle starred in the emerging auto art movement whose chief exponent was the aptly named Harrod Blank. In books and films, he championed those who covered cars with glowing strings of lights or painted them in exotic patterns. In Blank's film *Art Cars,* an amazingly high number of the decorated cars were Bugs. Bugs were painted and gilded, covered with sequins and sparkles, or tie-dyed. There was a Beetle whose wooden body had been carefully crafted by an Italian cabinetmaker. The body of another was replaced with cast iron latticework, making a kind of New Orleans Bug. There was even a live Bug, covered with grass growing from mesh embedded with soil.

The literary counterpoint to the art car Bug was found in the books of

British novelist Geoff Nicholson. He turned the Bug into a new kind of literary character—not a personality, like Herbie, but an icon that drives the behavior of the characters. "I am interested in VWs the way Melville was interested in whales," he said. For all his characters the Bug was a key prop in a wacky psychodrama.

Nicholson first touched on the Volkswagen theme in his novel *Street Sleeper* in 1987. But in *Still Life with Volkswagens,* he presents not just a cast of VWs but a cast of VW obsessives, from Fat Les, who restores damaged Bug bodies, to businessman Charles Lederer, who comes to believe Beetles are the devil's car. Mystic "Zen Road Warrior" Barry Osgathorpe has retired from the highway and become a proponent of the "green Bug," an ideal Bug that never moves and therefore never burns fossil fuel. His all-black Volkswagen, named *Enlightenment,* sits idle.

But the key character is a wealthy collector named Carlton Bax, who has a huge collection of Beetle memorabilia and Beetle-shaped products. He brushes his teeth with a Beetle toothbrush, showers with Beetle-shaped soap. The punch line, of course, is that he actually drives a Range Rover.

In using the Beetle as an icon to play off his character's obsessions and emotions, Nicholson implicitly showed that he understood the degree to which the Bug had become a blank, a cipher. The Beetle was now patently an object of nostalgia, a thing from the past, ready to be laden with new meanings. Nicholson supplied one of the clearest explications of the status of the Bug's appeal after it was no longer sold in Europe and the United States. He has a character explain:

> It's not pretty in the ordinary sense but it has a basic simplicity that is very appealing. Even people who found it downright ugly will admit that its eccentricity is part of its charm. Visually it strikes a strange balance between the reassuring and the sinister. . . . Because it's so ubiquitous . . . it's easy for people to stamp their personality on. In lots of ways the Beetle is like a kit car; bits can be removed and added quite easily. It's relatively simple to customize and modify it to make it your own. It has real character in a world where consumer products are increasingly dreary and empty. That counts for a lot.

Another character praises the Bug's "power, an elemental, archetypal quality, the sense of sculptural integrity."

Nicholson did not have to exaggerate the facts about real-life Beetle buffs very much to turn his story into dark comedy. Carlton Bax has collected historical items, the rarest and most memorable of which is a Nazi-era Beetle that goes only one step beyond the model Porsche created for Hitler's birthday: a pornographic mechanical model with interacting figures of Hitler and Eva Braun.*

These imaginative takes on the Bug were only extreme versions of real doppelgängers that had followed the car since the beginning.

The cabriolet model in which Ferry Porsche drove Hitler away from the 1938 cornerstone ceremony may be considered the first variant. Its supercharged 50 horsepower engine was twice as powerful as the stock model. Porsche was secretly experimenting with other models built on the Beetle basics, including the Type 64, the vehicle devised for the Berlin-to-Rome race that was canceled by the war. Its streamlined body, which resembled those of the Auto Union racers, was mounted on a Volkswagen chassis. And very quietly, because such diversions were not approved by the regime, the sports Model 114, its body designed by Erwin Komenda, proceeded far enough to undergo wind tunnel testing in 1939. After the war came the Hebmüller convertibles, whose small numbers make them valued by collectors today, and then the mass-produced Karmann convertible, the realization of the Beetle cabriolet promised before the war.

The Beetle chassis was adapted to other very different bodies. As convertible or sports coupé, the Karmann Ghia, named for the German coachworks and the Italian design house that jointly produced it, arrived in the late 1950s. Legends cloud its origins. American designer Virgil Exner, who later created the tail fin "Forward Look" for Chrysler, seems

---

* Nicholson has morphed this model with the many other over-the-top birthday presents Hitler actually received such as a model of the house where he grew up, the gift of the Krupps steelworks, made of the same Enduro KA-2 steel alloy used in armor plating and shells—and, incidentally, to roof the pinnacle of the Chrysler Building in New York City.

to have had a hand in its design, and the influence of the prewar Model 114 is also evident.

The formal name of the Beetle had never been anything other than Type 1; the VW bus was Type 2. It was the inspiration of Ben Pon, the Dutch importer who knew what small business customers needed and who sketched it out in a notebook in 1947. Three years later production began of the classic bus, also known in West Germany as the "Bulli" and around the world as the Combi. It became especially popular in the Third World where it provided basic transportation for everyone from missionaries to mercenaries.

In the U.S., the bus became a sort of generic transportation, the simplest cheapest box to move more people than the Bug. Self-described members of the counterculture drove the Bug and the bus with the same attitude with which they wore old army jackets and carried army backpacks and gas mask bags with peace signs drawn on them in Magic Marker. They were like guerrillas who had captured the weapons and matériel of the enemy. After all, they believed they were living in the heart of the beast, Amerika, a fascist regime.

The bus soon became a hippie cliché: a vehicle like a loaf of bread, packed with bodies on the way to a Grateful Dead concert. It was a miniature mass market version of Ken Kesey's school bus, with the simple destination "Furthur" painted on its front.

When Jerry Garcia died in 1995, VW ran an advertisement that simply showed the sketched front of a bus with a tear running down its front. It was the old bus, too, and not the newer, more rationalized model.

The people's car ideal held out the same appeal in other European countries that it did in Germany before World War II.

In Italy, even before World War I, Fiat began to search for a minimal people's car, not for a whole family but a couple and luggage. In 1913 it built a 500cc model, a 1918 experiment had a 760cc engine, and more studies followed. But it wasn't until Dante Giacosa, an aeronautical engineer, got involved that a 500cc model was introduced in 1934 as "the little

car for work and for thrift." It would be turned out by the thousands at
Fiat's vast Mirafiori plant in Turin.

Giacosa cleverly placed the radiator behind the engine, a miniaturized
four-cylinder, and strengthened the front of the car by using the engine as
a structural element. The Fiat's leaf spring and shock absorber suspension
were more traditional than the Bug's torsion bar arrangement.

Formally the Fiat 500—its engine had a capacity of 570cc—it went
into production in 1936. The shapes of its headlights suggest Mickey
Mouse ears and it quickly became popular as the Topolino—Little Mouse
or Mickey Mouse. When production ceased in 1948, half a million had
been built. The original Topolino and its successors such as the updated
500 of 1957 evoked the same respect and affection as the Bug.

The French 2CV, soon nicknamed the Deux Chevaux, looked like a
cruder descendant of Le Corbusier's Voiture Maximum. It seemed to
have been designed for rural French life. The design brief for the 2CV
included the specification that it should be able to carry a basket of eggs
across a French field without breakage. Its rippled metal body was as
old-fashioned as the Ford Tri-motor or a Fokker airplane of the 1920s.
Introduced before World War II but not produced until after 1945, its
very name Two Horses was a pun not just on horsepower but on the
standard farm team. It was designed and assembled in flat pieces and
suggested some prefab farm outbuilding more than a modern mode of
transport and soon came to symbolize the ruggedness and perhaps the
stubbornness of the French peasant.

By contrast, the British Mini of 1959 combined the surprising technical innovation of British industry with a stylistic gawkiness that ultimately became lovable. Created almost single-handedly by the Cyprus-
born Alex Issigonis, the Mini innovatively mounted its front-end, water-
cooled engine transverse style, or parallel to the axle. The result was more
passenger space and simpler gearing. The engine and transmission
shared a single oil sump. The suspension consisted of little more than
rubber cones. It was of a piece with British design for telephones, radios,
and teapots—an artifact of postwar, shortage-ridden Britain—but stood
in contrast to the decline of the British auto industry in general.

At first popular because it was cheap, within a few years the Mini was as beloved as the Beetle and equally associated with the counterculture. As with the Beetle, there were contests to stuff more and more people into a single Mini, but it had more interior space than the Bug so its numbers were always larger. Mini meant Carnaby Street and Mary Quant, Twiggy and Ringo Starr. The company even fostered the myth that the miniskirt had been named for the Mini car. Like British music and fashion, it proved exportable. The new automotive type form of the Mini was adopted around the world for small cars, in the U.S. and Japan and ultimately in Wolfsburg itself.

The East German Trabant became yet another doppelgänger and in its evolving symbolism perhaps the strangest of all.

The Trabant was the Beetle's competition in the Cold War, the East German "people's car." It was homely and plain, a pitiful piece of engineering whose design was frozen in 1962, with a body of a kind of plastic, and East Germans had to wait years to buy one. The Trabie became a symbol for the common man and his endurance in the face of socialism. In a particularly German, wry way, the Trabant seemed to win affection for its very deficiencies much like the Beetle and other people's cars. Owners happy to have any car at all seemed to feel an almost protective liking for the car, as for an ugly but loyal dog. The Trabant suggested that it was not so much aesthetics or engineering that engendered affection for a car as cultural context. It was an expression of optimism despite the flaws of humanity and its political system.

The Trabant was also an escape vehicle. At the Checkpoint Charlie Museum in Berlin, named for the most famous crossing point of the Berlin Wall, is a room full of devices used to escape the East. Amid the aircraft and boats in which refugees were smuggled out sits a heroic Beetle. Nearby is an Isetta, produced by BMW in the mid-1950s, the most famous of the tiny bubble cars that represented the efforts of once noble automakers to stay alive in postwar Germany. It had a single door in front to which the steering gear was attached. It was so small that it was the only car that regularly escaped examination by East German border guards, who reasoned there was no way a person could be smuggled inside such a tiny vehicle.

In the Isetta on display, the original gas tank had been replaced with one holding a few ounces of fuel to make room for a human being. Beside it, the Isetta's single-cylinder engine was hung like a magic icon, a kind of metallic heart, in a vitrine. It is testimony to the economic difficulties of BMW as well as to the cleverness of the East Germans.

But most prominent among the vehicles was a colorful painted Trabant, the East German people's car.

As one observer shrewdly put it, that such a car had been produced by German engineers and workers was the most resounding condemnation possible of the communist system. Like the Bug and the Model T before it the Trabie was the butt of jokes. A Trabant flew off the road and came to rest in a cow pasture. "What are you?" asked the meadow muffin beside it. "I'm a car," said the Trabant. "Yeah," replied the meadow muffin, "and I'm a pizza."

The Trabant's competitor in the East was the Wartburg, a car as ugly as its name and even more technically disdained than the Trabant. The Wartburg was named for the town of Wartburg near Eisenach, where it was produced in a former BMW plant, and the name suggested an unintentional parody of Wolfsburg.

The Trabie and the Wartburg could be viewed as the Beetle's competition—the best the East could put up against the star product of the West German automobile industry. Cartoonish Trabies painted on the Berlin Wall appeared to smash through its masonry, the mild-mannered weakling of a car miraculously transformed into a superman. But after the Wall came down, in November 1989, the painted Trabants were replaced by real Trabants driving into the prosperous, wondrous West.

chapter thirteen

f or Carl Hahn the Trabant was a symbol of all that had been wrong with the East. It was a shame, he told a *New York Times* reporter in 1990, "that East Germans had been forced to wait fifteen years for a car that was below human dignity."

Reunification changed the way Volkswagen saw the world. After 1989 Wolfsburg was no longer on the front lines of the Cold War. Hahn declared that after the fall of the Wall, Wolfsburg was suddenly in the center of Europe, equidistant from London and Moscow. "The way looks clear to Moscow," he declared.

Hahn, the son of an executive with the prewar Auto Union, had been born in Chemnitz. Hahn had hired DDB as the company's American ad agency in 1959 and had led the growth of the company in the U.S. market from 1959 to 1964. While there, he had married an American; his wife, Marisa, was the sister of the romance novelist Danielle Steel. In the 1970s, when Hahn was passed over for VW's CEO, he left the company and became chief executive of the ailing Continental Tire Company. While he turned Continental around in the late 1970s, VW slumped. He returned to Volkswagen in triumph in 1982 and presided over the company's revival in Europe. He was regarded in West Germany as a cosmopolitan figure, a supporter of the arts, well dressed, fluent in several languages.

Hahn began to define Volkswagen as a global company. He signed agreements with Toyota and Nissan in South America and Asia and opened a joint operation in China in 1983. He hoped to retire as a sort of economic statesman.

Even before reunification formally took place in October 1990, VW and other major West German companies moved to expand into the East. Hahn worked to revitalize the East German industry with a $7 billion investment.

The ultimate victory of the Beetle over its communist cousin came when Volkswagen took over administration of the old Trabant plant in the spring of 1990 and replaced the East German engine with a VW one. At the old Trabant factory in Mosel, the three millionth Trabie was ceremonially chosen to be the first with a VW engine.

By July, a whole plant in Chemnitz—formerly Karl Marx Stadt—was shifted to production of VW engines. In September, Hahn joined Chancellor Helmut Kohl in Mosel to lay the foundation stone for a new, modern factory. (Opel also opened a new factory in Eisenach to replace the Wartburg plant.) VW also took over the Czech firm of Škoda in 1991.

The benefits of the West were slow to arrive, and the Trabie became a symbol of the East German resentment of the arrogance of the West. It took workers a week to earn what Wolfsburg's workers earned in a day, and the pay at Škoda was even worse.

When the Wall came down, cynics predicted that West Germans would soon be eager to build it back. In one sense, they were right. Many West Germans resented the costs of unification, the influx of refugees and their effect on the economy. The East Germans, or "Ossis," in turn, resented their treatment as second-class citizens and felt their sufferings and struggles were unappreciated. They grew nostalgic for the socialist days, grim as they were, recalling a time of employment, controlled prices, and safe streets. The affection for the Trabant became part of a wider "ostalgia," which reached the point of absurdity when one man proposed to build a DDR theme park outside Berlin, complete with watchtowers and barbed wire, ration coupons and bread lines, surly clerks and secret informants.

The Trabant was the automotive equivalent of the Ampelmann, or

Lamp Man, the Mr. Magoo–like figure who had presided over pedestrian crosswalks in East Germany. The Lamp Man (green for go, spread arms and red for stop) was designed by an East German traffic psychologist named Karl Peglau in 1961—just as the Wall was going up. East Germans had the Ampelmann drilled into their heads from kindergarten on; he was as primal an image for them as Smokey Bear is for Americans. After unification, he was replaced by the standardized pedestrian signal used throughout Europe—Euromann defeated Ampelmann.

But Westerners as well as Easterners rallied to save him, and a number of crossing lights using the Ampelmann were preserved. Marcus Heckhausen, a West Berlin graphic designer, even organized a society to preserve the Ampelmann figures. He edited a book celebrating them and began selling Lamp Man novelty items—T-shirts and key chains—to support the cause. Affection for it and the Trabie and other objects was evidence of the desire to retain national differences in the face of a unified Germany and the integration of Europe. Defending Ampelmann was like defending local apples against the Euro fruit standard or regional dishes against McDonald's. The irony was that a relic from an inhumane regime seemed more human than the standardized symbols of capitalism.

The West came to resent the East, too. The Ossis were like poor relatives who had moved into the spare bedroom. As the costs of reunification rose, West Germans also felt a mixture of resentment and nostalgia.

In Wolfsburg, which had treated Italian and Turkish *Gastarbeiter* with disdain in the 1950s and 1960s, resentment of the Ossis was especially strong. When a sociologist named Martin Schwonke studied the town in the 1990s he found a city full of seething conflicts, where the class and ethnic differences of the factory pervaded all of life. Only 40 percent of people in Wolfsburg, he found, felt good about their neighbors. People were also nostalgic for the Cold War's shared sense of purpose and the common enemy, which had plastered over differences in class and nationality.

Within a couple of years of unification, the unprecedented costs of absorbing the sluggard East German economy threw Germany into an economic slump. The government had been too optimistic about the speed with which the East could be integrated and modernized. Volkswagen's situation was a microcosm of the country's: it was overextended. In trying to bring the East's factories up to Western levels of quality and efficiency, it had taken on too much too soon. Worldwide, profits fell from some 3 billion deutsche marks in 1989 to 1.78 billion in 1991.

But the recession revealed something else. The efforts to repair the damages of socialism highlighted the problems of capitalism. For all its robots, Volkswagen's cost structure in West Germany made it uncompetitive. Its fixed plant and labor costs were so high that even at full production at Wolfsburg, the company could not make money. Ironically, VW was a quasi-socialist enterprise itself, 40 percent owned by the federal and state governments. It lacked the flexibility of purely private enterprise while also failing to enjoy the full support of complete government ownership.

A new view of the factory was reflected in a 1989 film by the avant-garde documentary filmmaker Hartmut Bitomsky, *Der VW Komplex*. Bitomsky moved beyond the bright images of heroic capitalism and grateful workers to a more skeptical attitude about the assembly line. The robots were noisy and vaguely sinister. Bitomsky intercut scenes of the factory with those of crowds and empty damp squares in Wolfsburg. His ponderous camera lingered over the barriers between workers and robots. Workers were seen as alienated, mere attendants to robots. Bitomsky contrasted the noise and darkness of the modern factory with the quieter, brighter one of the 1950s. He juxtaposed computer control panels with the simple images of Peter Keetman's 1953 black and white photos and intercut the cheerful crowds of workers in the film *From Our Own Strength*, also made in 1953, with the dull masses of Wolfsburg in the 1980s.*

* John Updike, in his experiment in magical realism, the 1994 novel *Brazil*, sketches a similar vision of alienated workers subordinated to machines. One of the characters works in a VW factory. "The Volkswagen was a great-hearted machine . . . designed by a famous sorcerer called Hitler to take the German masses to a place called Valhalla," the hero is mockingly told. "The cars, little Volkswagen 'Beetles' painted the shades of tan and brown that

In many ways the increasing impersonality of VW's factories was the legacy of the company's long dependence on the Beetle, and that dependence continued abroad. VW was still essentially a one-car company in an era when Japanese lean production had supplanted the Fordist model. Factory wages had grown so long as demand increased for the Beetle and for the later Golf and more efficient ways were devised for building them. When the Golf was conceived as a successor to the Beetle, production and engineering dominated VW's thinking at the expense of design and marketing. If, as the saying went, the Beetle had "sold itself," Volkswagen never learned to respond to buyers' needs.

Unfavorable exchange rates and neglect by German management, who had taken the U.S. for granted, meant that VW's situation was even worse in America than in Europe. Post-Beetle VW suffered from bad design, along with poor ratings in consumer surveys for reliability and service, and carried higher prices than Japanese small cars.

Even with Hahn, the former American boss, at the head of VW in Wolfsburg, German executives had remained deaf to the requests of dealers and customers in the U.S. The Passat, for instance, Volkswagen's largest model, was a good automobile, a four-cylinder bread-and-butter family sedan with the potential to compete with the Toyota Camry or Honda Accord. But American dealers and salesmen had argued that to be competitive the Passat needed a six-cylinder engine and luxury features like power windows.

VW's design had been declining for years. The Rabbits and other cars that came out of VW's Westmoreland, Pennsylvania, plant, run mostly by former General Motors executives, struck the wrong note. In some ways the cars were too German, in others too American. Volkswagens of the time had angular dashboards, tiny wheels, and out-of-fashion colors and materials. Placement of controls and seat adjustments struck Americans as eccentric, and the fake wood grain and Detroit colors that

---

gave them the name *fusca* in Brazil, were manufactured in a giant shed whose northern end, like a hungry mouth, took in Volkswagen parts and whose southern end, like a tireless anus, emitted the completed *fuscas*." Updike's hero is assigned the task of sweeping up around the machines where "the racket of assembly was so incessant and loud . . . the machines made machines of men."

replaced the dark and plain interiors of the European models ended up alienating the traditional VW buyer and his slightly snobbish preference for a "Euro" look over a Detroit one.

In the 1980s in the U.S., the post-Beetle Volkswagen presented itself as the company of "German engineering." Mercedes and BMW had German engineering, too, of course; what was implicit was the VW's affordability. But when competing with the Japanese, who held a price and quality advantage and whose engineering, if less exciting, was at least equal to VW's, the pitch fell flat.

The Japanese had carefully studied the tastes of U.S. customers and had integrated what were known in the industry as "surprise and delight" features into their cars. They were little things that were symbolically powerful. The most famous was the coin holder that Toyota and Honda offered. Drivers appreciated the gesture and did not feel the same way about Volkswagens.

In 1988, the Westmoreland VW factory was closed—just as the Japanese were expanding their U.S. plants. In 1990, VW sold fewer than 100,000 cars in the U.S., a third the number of two years earlier. Durable and inexpensive Japanese cars continued to expand their share of the American market, and Detroit, which had finally begun to study Japanese techniques, had a resurgence. Chrysler, on the brink of bankruptcy in 1978, came back a few years later with a new generation of cars. In 1985 Ford was saved by the Taurus.

Paul Lienert, one of the auto business's shrewdest observers, diagnosed VW's problems in a prescient 1991 article in *Automobile* magazine. Volkswagen was competing not with Detroit but with the Japanese, he argued. Its edge lay in technical innovation—the nuts and bolts of "German engineering," so to be successful, it should emphasize those technical innovations. And its design had to improve. Lienert wrote, "VW's design efforts could use a massive injection of chutzpah—maybe even a little whimsy. The company simply can't afford to prolong its staid image and conservative styling in today's frenzied, fashion conscious environment. More frequent styling changes, a grander vision and a renewed effort to regain the lead in design could help set VW products apart once again."

Lienert commented on the "inexplicable myopia" of VW's German executives. In the wake of reunification, Hahn and others increasingly saw all local markets as subsidiary to Europe and the goal of building a "world car"—one model with local variants produced around the world. Few cars could be called world cars—the Beetle was one.

In an effort to bring some wit to the "German engineering" pitch, VW launched its Fahrvergnügen ad campaign. Fahrvergnügen was an invented German word that ostensibly meant driving pleasure. The ads were widely derided and lent themselves to an amazing variety of parody, much of it off-color. But the real problem lay in the cars themselves, which, for the most part, were joyless and cold.

A few models such as the sporty GTI—introduced with an ad that substituted German voices singing "little GTI" in the place of "little G.T.O." from Ronny and the Daytonas' 1964 hit "G.T.O."—captured the idea and garnered good sales.

The German engineering theme, in a more specific and tangible form, had worked in a different way for Audi, VW's premium brand. In the early 1980s Audi's U.S. sales soared on the strength of aerodynamically sleek shapes—five years before Ford's landmark "aero" Taurus—and the four-wheel-drive system, trademarked Quattro. VW's message was less clear.

Volkswagen also seemed plagued by plain bad luck. A new chief of VW in the United States, James Fuller, began to repair relations with dealers antagonized by German policies and market Volkswagen's technical heritage. But on December 21, 1988, Fuller and Jim Marengo, another top VW of America executive, were killed in the terrorist bombing of Pan Am Flight 103.

After Fuller's death, Volkswagen began marketing itself as the "honest value" company. VW, the pitch went, was the car that gave you a lot for your money. But its cars were more expensive than those of the Japanese, whose dependability and quality were rated higher. These values mattered to the customers, who, although they might have bought Beetles in the early 1970s as first cars, now had jobs to get to and children to worry about.

Carl Hahn, the man who had found America for the Beetle, was in danger of becoming the man who lost it. He was taking the blame, but he had taken steps to improve things that would take time to pay off. One was the establishment of a center in California to track trends and create designs for the American market, although VW opened its Simi Valley studio a full decade after Japanese car companies had established similar satellite studios.

When Toyota, Honda, and Nissan began to sell cars in the U.S. in the late 1950s, they made cultural mistakes trying to speak the American car language. Some of their first cars bore silly names, such as Nissan Fairlady—inspired by the Broadway show *My Fair Lady,* which they believed would appeal to American buyers. The shapes and colors of the car were equally gawky. The satellite studios were created to correct such mistakes. The companies chose to set them up in California because the state provided American automobile culture in its most concentrated and advanced form. Soon, Detroit companies followed. Now it was Volkswagen's turn to acknowledge that it was out of touch with the U.S.

chapter fourteen

t he Audi Volkswagen studio, which opened in 1991, looked like hundreds of other speculative office buildings in the rolling hills north of Los Angeles. It was in a rapidly developing area where high-tech companies and luxury condos were taking over ranches and farms. Only a short drive from open highway, it stood among low office buildings and small malls with coffee chains and bookstores. Not far away was a new tourist attraction, the Ronald Reagan Presidential Library. Generously fenestrated with semiopaque glass, and meticulously landscaped, the studio could have been a small insurance company or an engineering firm. But at lunchtime and after work, passers-by might see radio-controlled model cars screaming around the parking lot.

The story of the design of the New Beetle was established as legend almost from the beginning. The mythic moment of creation was captured in a painting commissioned by VW by John Marsh. It was like John Trumbull's painting of the signing of the Declaration of Independence, or David's portrayal of the Tennis Court Oath—a microthin slice of history frozen into poses that were expressive but wholly artificial.

The painting also echoed that classic genre, the artist's studio. It showed the Simi Valley studio full of patrons and assistants and surrounded by a still life of significant artifacts: a photograph of old Ferdinand Porsche himself; the small-scale wind tunnel model; a 1948 Beetle brought in for inspiration; the drawings for an alternative shape with fat pontoon fenders; a stray logo design or a hubcap.

Fifteen designers were employed at the studio, which would serve both VW and Audi. Their boss was J Mays, a native of Oklahoma, who had just arrived from Germany wrapped in the triumph of a show car he had designed for Audi called the Avus, named after the Berlin race track where Hitler and Porsche had first met. And he was skeptical about what California could teach Germans about cars and their design. He considered the design of European cars superior to the Super Bees and Turnpike Cruisers he had known as a child.

Mays and others at Audi had begun to look at the way history could inform the design identity of the company. The Wall had just come down, unification was in sight, and for the first time in fifty years, German history, even automotive history, was safe territory to explore. The Avus, Mays said, was the car "that taught me about design as communication." The echoes of cars of the past endowed the Avus with a meaning beyond its shape. It was a futuristic car that looked back: with a shape reminiscent of Porsche's 1930s streamlined racers, and crafted from aluminum, Audi's trademark material, it was polished to a mirror finish, like a computer special effect from the film *The Terminator*.

The legendary scene frozen in John Marsh's painting began at a lunch one day in 1991 at which Mays speculated about Volkswagen's dreary state with Peter Schreyer, a designer from Audi on exchange to the California studio. Customer surveys still linked Volkswagen to the Beetle. As fewer and fewer Beetles were seen on the road, their fans began to remember them as an ideal, an abstraction, almost an archetype.

A new Beetle was the most obvious idea in the world. If asked, How do you fix VW?, anyone on the street would have told you: bring back the Beetle. The problem was the idea's very obviousness. But Mays discussed it with another designer, Freeman Thomas, and the more they talked the more excited they got; they began to think of themselves as conspirators with a secret plan.

Mays, Schreyer, and Thomas concluded that the company had "walked away from its customers." They isolated what Americans had liked about the company—the Beetle and the values it embodied. Mays counted off the key adjectives on his fingers: simple, honest, reliable, fun.

They created a design language to express these values. The circle, Mays explained, was the most honest shape. It would become the key to a new Beetle-shaped car they called the Concept One.

They broke the shape of the old car down into its essential geometry, reducing the car to three arcs—the two fenders and the roof—then inverted those arcs in the smiling lines of the hood and trunk. They rendered not just the essence of the actual car but the memory of and appreciation for an American icon. It was a shape as simple as the logo Mays drew on his Macintosh, three arcs, three circles. It was as recognizable as Mickey Mouse.*

From the earliest sketches, the Concept One was cartoony and toy-like. The car was drawn in yellow—smile button yellow, McDonald's Happy Meal yellow. Strength through joy indeed: the Bug had gone from small car to smile car.

Mays (his middle name was Carroll; the J stood for nothing except itself) was born in 1955 in Maysville, a tiny town in south-central Oklahoma that bore the family's name and where his father ran an auto parts store. To hear him tell it, it was a kind of *Last Picture Show* town. "We sat around the loading dock and gas pumps on Saturday night waiting for something to happen and of course it never did." Discussion would turn to such burning topics as the virtues of Ford versus Chevrolet engines.

*Alluding to the simple shape of Mickey Mouse was already an established gesture of high design. Ettore Sottsass, the Italian designer behind the Memphis movement, had reflected Mickey in an office chair. The American architect and designer Michael Graves adapted Mickey ears in a pepper grinder for the Italian firm Alessi, and the Japanese designer Kita evoked them in his 1980s Wink chair.

For excitement, there were trips to Pauls Valley, about twelve miles away, a town of such wider cultural opportunities that it had a drive-in theater and a Dairy Queen.

Mays worked in his father's store learning the nuts and bolts of automotive technology. The job lent him a detailed understanding of the difference between European and American technology that was visible in even the most mundane parts. "In Germany," he remembered, "you can find seven different kinds of exquisitely designed light switches, all expensive. In the U.S. you find one basic light switch, for $.79." He bought an offbeat European Ford Taunus and in his spare time sat on the roof of his friend's purple Super Bee, a Dodge muscle car whose color must at some point have matched a hue in the constantly changing Oklahoma sunset, listening to Merle Haggard on the radio and dreaming of distant worlds.

One small-town ritual he particularly enjoyed came every September when the new models arrived at the local dealerships hidden behind covered windows until the morning when the salesmen pulled the paper off the glass.

Mays studied journalism at the University of Oklahoma but was drawn more to the power of the image than the word and soon transferred to Art Center College of Design in Pasadena, California. Art Center, the leading school for American auto designers, was housed in a glass and steel building that was flung like a bridge over a canyon by architect Craig Ellwood, a disciple of Mies van der Rohe. Art Center was just a few miles from Caltech and the landmark arts and crafts houses of Greene and Greene and seemed to imbibe some of the spirit of each.

At Art Center, teachers and fellow students told Mays "your stuff looks German." It was not a high compliment at a school where most of the faculty were veterans of Detroit design studios. But Mays was drawn to a pure modernist geometry his fellow students considered Teutonic or Bauhaus. He liked concepts that emerged fully formed and elemental, like an egg, needing no refinement or revision.

Fortunately, Mays was hired by BMW after he graduated in 1980. "My real education began in Germany," he recalled. He soon moved to Audi, where he worked under Hartmut Warkuss and learned the basics of designing a car: stance and proportion and the importance of detail. Warkuss also introduced him to the ideas of the Bauhaus and school of Ulm, its more commercial successor. Warkuss would say, "We need a little bit of Braun here," referring to the legendary manufacturer whose classic radios and coffeemakers were designed by Hans Gugelot and Dieter Rams. And Mays came to understand that while automobile ads told people "You are what you drive," the truth was that people drove what they wanted to be.

Freeman Thomas had graduated from Art Center a few years after Mays, although the two never met there, and had gone to work for Porsche, then run his own consultancy in California before he agreed to come to VW in 1990. Mays had discovered Europe; Thomas had grown up there. He was a child of the Cold War. The son of an American military man— an air traffic controller—and German mother, he was born in the U.S. in 1959 but spent much of his childhood in Europe where his father was stationed. As a child, he rode with his parents in a variety of cars—sometimes a small Mercedes 190, sometimes a large Buick Roadmaster. From these drives, Thomas had developed an intuitive sense about the interior spaces of automobiles.

Thomas grew up obsessed with cars and their history, but for all his encyclopedic knowledge of specific models, he was drawn to a certain purity of concept. He loved Porsches, but also such basic and timeless designs as the American rural mailbox, the Weber grill, or the classic yellow school bus.

After Art Center, Thomas got a job at Porsche, where he found himself spending months designing a gas filler cap or wheel center. Porsche design delighted him, but the company changed only slowly and offered designers few opportunities.

In Europe, both Thomas and Mays discovered a distilled modernism, the version of Bauhaus that the design school of Ulm had popularized during the 1950s. It was the look Dieter Rams and Hans Gugelot brought to appliances, such as Rams's famed compact stereo set critics nicknamed "Snow White's coffin," and the coffeemakers by Braun and Krups that became kitchen counter icons of "Euro" in the United States.

German executives isolated in Wolfsburg or Stuttgart or Munich sometimes failed to understand the American culture that nonetheless fascinated them. Mays and Thomas were in a position to interpret between the two. They made the Bug idea a personal campaign as only Americans would have done, and devised not just a car, but a whole new strategy for VW. Thomas declared they felt like "curators" of the ideal and history. It was a phrase Freeman Thomas had heard from his old boss, Harm Lagaay, the chief designer at Porsche, where the continuity of shapes, especially in the 911 model, was paramount. They would channel the soul of the Beetle.

They understood quickly that if they expressed the idea verbally it would be rejected immediately. Volkswagen executives had spent years trying to put the Beetle behind them, to abandon their dependence on a single model. They would have to show, not tell.

The Bug would have to be reborn by the power of its appearance. It would change again in image, not words.

Mays also understood how cars took on different meanings in different environments and contexts. Unlike the modernists, who believed in objects that radiated purity from some essential set of shapes, he knew designs changed in meaning as they moved from culture to culture—as the Beetle had. It was the culture that surrounded a car, Mays intuited, that made it mean what it meant and feel like it felt. "There is a lot of difference between a Mustang sitting in a dealership and a Mustang driven by Steve McQueen in *Bullitt*," he explained.

A group of outside consultants shared this understanding of context. Left out of Marsh's commemorative painting were the people from SHR Perceptual Management, a "visual positioning group." They brought in piles of research on Beetle culture and history, supplying the historical context Mays wanted, and contributed a visual language for the car, helping the designers translate their adjectives (simple, reliable, honest, original) into shapes based on circles and arcs.

SHR had been founded to advise companies on what it called "perceptual management." It pushed for "visual positioning," its phrase for assuring that the design of products and other visual elements sent out a consistent message. This was not the language of Platonic geometry or the Bauhaus, but of sales.

SHR's founders, Will Rodgers and Barry Shepherd, believed that as products grew more alike in quality and performance, design would be the key difference in sales. SHR helped sell the idea. The firm conducted exercises with all its clients that involved showing them sample colors, textures, and photographs. Employees were asked to select the samples that defined their image of their company or product. Was it rough or smooth, red or blue? SHR's method offered a kind of visual psychoanalysis of a company. It also helped sell design to a company's employees, because the sample selection involved them in the process. With SHR's involvement, it was clear that not only the Beetle would be reborn, but the whole VW brand. The Bug would rule.

Concept cars are built to attract public attention at auto shows and improve a company's public image. In the 1950s, General Motors began to show futuristic "dream cars" in its Motorama shows, but they were almost never turned into production versions. By the 1990s, it had become incumbent on companies to unveil either an important new model or a dramatic show car at the major shows—Geneva, Frankfurt, Detroit, Tokyo. Generally, concept cars were built around the conceit of new technology, which was often unrealistically expensive or immature and just a pretext for the style that was their real point. To provide such a pretext for the Concept One, the Simi Valley group decided that the car would accommodate any of three different "environmentally friendly" engines: an electric motor or a conventional gasoline one or a hybrid of both.

The California Air Resources Board, or CARB, had decreed that a certain percentage of each manufacturer's vehicles sold in the state, rising from 2 percent to 10 percent with each year, would have to be "zero emission" or electric vehicles. Manufacturers were scrambling to meet this requirement while simultaneously lobbying against it as impossible.*

One of the faults of the old Beetle had been its inability to meet anti-pollution standards. The technological rationale for the new Beetle would be its low emissions; the public relations rationale—and the argument put to higher-ups—was that the car would demonstrate Volkswagen's efforts to comply with the CARB rules and its sensitivity to the environment. In memos Mays and Thomas code-named the car "lightning bug."

Other makers were looking backward for the shapes of their cars, trying to stand out among vehicles that were increasingly perceived to have grown to look too much alike, in part because of the aerodynamic shaping that began in the mid-1980s.

The hottest car of the time was the Miata, which had come out of Mazda's California studio in Irvine, an hour's drive from Simi Valley. It was a rear-wheel-drive sports car with a softly shaped body that seemed to sum up everyone's memories of British sports cars, of the MGs, Austin Healeys, and Triumphs that showed up in America around the same time the Bug had.

The Miata. It was immediately called retro, but it did not take the shape of any specific older model, rather it was a distillation of their shared gestalt, producing a sense of familiarity. "Haven't we met somewhere before?" was the ad slogan. In the background were faded posters for old MGs. One of the car's ads said it best: "Zero to the Sixties in seconds."

The Miata appeared almost fetal in its soft shapes and hidden headlights. The front turn lights suggested the eye buds of a fetus, the taillights were an abstract combination of blobs that led curator Paola Antonelli to include it in a show of Mutant Materials at the Museum of Modern Art in

---

* The board was rigid in its view: "zero emission," strictly speaking, meant only electric cars, which were far less practical than "ultra-low-emission" cars that used a hybrid of internal combustion and electric motors.

1995. The Miata was not only cleverly designed,* it was well made. Japanese automaking supplied two things the British originals resoundingly lacked: quality construction and mechanical reliability.

The Miata appeared at a time when companies were looking to what marketing experts call "relationship" marketing and "aspirational" design. Saab, introducing a new model, held showcase picnics and presentations for its owners, assuring them that the ignition key was still on the floor, and that the car hadn't lost its lovable eccentricity, even though the company was now owned by General Motors. The idea was to bond customers to the company as to a club, to lend some of the great sense of membership, community, and belonging that Beetle owners had long felt—to give them a car they wanted, not just needed, and to make the purchase something more than a cold hard cash transaction.

On such premises Chrysler launched the Neon in 1993 as the company tried to figure out how to successfully sell small cars. The Neon would attempt to create the same sort of appeal for a small car that the Beetle had enjoyed, and revive the market for small cars, which most U.S. makers had conceded to the Japanese, by giving the car more personality and character. Chrysler's planners explicitly looked to great ad campaigns: the Pepsi generation; Miller Lite Beer; the Apple Macintosh; and, of course, the Volkswagen Beetle.

The Beetle's influence on the Neon was clear in its round headlights and smiling front. The slogan "Say Hello to Neon" was clearly borrowed from Apple's "Hello, I'm Macintosh" screen greeting. Both products presented themselves as little pals. "Cars are like friends" was another line in the Neon ads.

Chrysler intended to bind Neon owners and buyers in a clublike relationship. Buying and owning a car, the marketing gurus were now saying, should be a process analogous to joining the club—"affinity marketing," the business schools called it.

* Some of the car's fans saw a neat symbolism in the fact that one of its designers was Mark Jordan, son of Chuck Jordan, GM's head of design and the man most often blamed for the bloated Chevy Caprice, the Shamu-like family car that marked one of the low points in the history of American car design.

Volkswagen, however, the textbook case of "affinity marketing," had reached a low point. In 1993, sales would fall below 50,000 cars. The once proud network of white box dealerships with blue signs was tattered and gray. At Gensinger Volkswagen in New Jersey, salesmen spent most of their time selling used cars or reading the sports pages. They had few new cars. Most were Golfs and Jettas made in Mexico, where quality had declined so badly that most of those they received from the factory were shipped back. By 1993, the Wolfsburg executives were discussing pulling out of the U.S. market entirely, as Renault, Fiat, and Peugeot had.

The Simi Valley designers were supposed to be engaged in gathering information on colors and materials popular in the U.S. and preparing designs for future models to compete with those developed in the German studios. But Mays and Thomas's idea was nothing less than a strategy to re-create VW in the U.S. market.

In the summer of 1992 they decided they had to trust someone within the company. Mays flew to Ingolstadt, Germany, to meet with Hartmut Warkuss, his former boss from Audi design, and shared their secret. Bemused by the Americans' initiative, Warkuss nonetheless gave the go-ahead to a $300,000 budget to turn the sketches and small models into more refined, quarter-scale models. Warkuss ordered the project be kept secret, a Skunk Works–style operation hidden from other designers in Simi Valley. He, too, realized that not everyone would understand the idea of a new Beetle and that it could be sold only through finished models. There was nothing less likely to appeal to VW's American executives than a revival of the Beetle. As Beetles grew fewer and more battered on the American highway, they seemed a reproach to the company: what else had it ever done?

Mays and Thomas, along with skilled modelers Dave Morris and Richard Woodley, worked out their ideas in detailed one-quarter models around two basic themes. One, developed by Mays's team, was a more geometric and sober Bug, the other, developed by Thomas's group, was more flowing, whimsical, and nostalgic.

Mays and Thomas realized that if they played the politics right, the attitude of Volkswagen executives might change. Their efforts received a sudden impetus in February 1992. The Vorsitz, or board, of VW, concerned by the company's falling sales and poor financial situation, decided to oust Carl Hahn. The man who had built up the Beetle in the U.S. and hired DDB was cast aside. He would retire early and in January 1993 be replaced by Ferdinand Piëch, the chief of Audi.

Audi was one of the few bright spots in the financial picture of the larger VW organization. Since Piëch had become its head in 1988, he had built the company into a rival of BMW and Mercedes. *"Vorsprung durch Technik"* was its motto. "Progress through Technology" is the common translation, but *Technik* implied craft and engineering, joined together. Piëch had helped turn Audi, which VW bought in 1965, from a stodgy company whose models were driven by bureaucrats, into one that made exciting and technically innovative models.

Piëch was the son of Louise Porsche and Anton Piëch, who managed the KdF factory during World War II and died in 1952, and the grandson of Ferdinand Porsche. Piëch was born in 1937 and spent the war mostly in remote Zell am See, Austria. He studied engineering at the Swiss Federal Institute of Technology in Zurich and in 1962 took charge of the Porsche racing program, producing the landmark 908 and 917 models. Under his direction Porsche won Le Mans and several world championships.

He was a harsh boss. One story claimed that he had once stood outside the restrooms at Porsche with a stop watch. Many observers believed he was encouraged by his mother. Indeed, to the end of her life—she died in 1998 at age ninety-five—Piëch visited her every couple of weeks. At Zell am See and other family retreats, she went deer and bird hunting well into her nineties. She was not just an ambitious mother but a shrewd businesswoman. She obtained rights to the Austrian Porsche and Volkswagen sales organization and built up an empire of real estate worth some $50 billion, making her one of the wealthiest women in Austria and Germany. Piëch himself shared in the Porsche wealth; he was estimated in 1999 to be the richest man in the German auto industry, owner or executive, with a net worth of $5 billion.

In 1972 Piëch became caught up in a nasty rivalry with his cousin, F. A. Porsche. To resolve the conflict, Ferry Porsche, Mrs. Piëch, and other family members on the board exiled both men and decided that no family member should ever again head the company. Piëch went to Audi, where he rose through the ranks of research and engineering. He was a key force behind the landmark Audi 100, the aerodynamically shaped car. In 1982 he introduced the Quattro, the four-wheel-drive system that soon became the trademark of the company.

Audi was the last surviving part of the prewar Auto Union, the second largest German car company in 1939. But after the war most of the company's factories were located in East Germany and were dismantled by the Soviets. It had taken decades for Audi to rebuild itself.

Its streamlined bodies and European performance made Audi a sensible alternative to more expensive BMWs and Mercedeses in the U.S. But in 1986 the CBS television news magazine *60 Minutes* ran a segment charging that some Audis were subject to "unintentional acceleration" and had suddenly jumped into gear and pinned drivers against garage walls. The company denied the charges and handled the bad publicity poorly. Piëch had helped Audi recover from the episode. Audi celebrated its return to prominence—and implicitly the reunification of Germany—with J Mays's Avus concept car, shown at the Tokyo motor show in 1990.

When Piëch took over as head of VW in January 1993, he promoted Warkuss to replace Herbert Schaefer as head of design for Volkswagen. In July, Warkuss arrived in Simi Valley with two lieutenants to see what Mays and Thomas had produced. But before the Germans were allowed to see the detailed quarter-scale models of the car, along with rough full-sized clay ones, Thomas and Mays made them sit through a seven-minute presentation of images and sounds—the vital context. At the end of it the three visitors rapped their knuckles vigorously on the tabletop, a German version of applause.

Warkuss approved going forward with a full-scale model. The two basic variants—the geometric and the retro—were combined so that the circles of the headlights and the strong arch of the roof were counterbalanced by a more fluid, biomorphic, almost pneumatic set of fenders.

The final result, completed at the end of the summer of 1993, was an essay on the American love for the Beetle. It was no accident that the glowing green of the Concept One's dashboard came from the color of Mays's swimming pool, beside which the team had photographed the original prototype. The car was not so much a homage to the original Beetle as to the California dream of that Beetle—a Beach Boys, good vibrations California dream.

The details of the Concept One, especially in the Bauhaus style interior, were precise and mechanical. The seriousness of the engineering played off the whimsy of the basic concept. The combination of wit and deadpan precision produced an effect like a latter-day Pop Art experiment: What if you built a front-wheel-drive car in the shape of the old rear-wheel-drive Beetle? they asked, just as Claes Oldenburg had asked, What if you built an oversized typewriter out of soft cloth or a huge baseball mitt out of sculptural steel?

The transparent fiberglass model bore the name Concept One. (In VW literature the Beetle was called the Model One until 1968.) It bore the three-circle logo Mays had designed on his Macintosh.

In September of 1993, the model was shipped to the Valhalla, the highly secure viewing room at VW headquarters in Wolfsburg. The secret concept cars shown there were known as "Erlkönige"—elf kings, a word from a famous Goethe poem suggesting a sinister mystery. The model unaccountably went missing for a couple of weeks, in one version of the story, it was hidden by those who were appalled by it. But once Piëch saw it he approved. *"Alles in Ordnung,"* he pronounced. Everything in order—an engineer's praise.

In the last months of 1993, the group in Simi Valley worked out the final Concept One details. At the same time, they developed a dramatic design for the stand where the car would be presented at the Detroit auto show, and, working with SHR, conceived clever and evocative press materials and a multimedia display. They still did not trust the car to sell itself nor could they count on the German executives, who remained opposed to the idea, to sell it.

In January 1994, at the Detroit auto show, the Concept One was unveiled. Mays and the rest of the group were nervous. At the formal press conference, Ulrich Seiffert, VW's chief of research and development, the man behind the highly praised VR-6 engine, spoke first.

Before the car was rolled out, he declared flatly, "We will not be building this car." It was simply a project to demonstrate the company's commitment to the American market, he said, and to evoke the emotions of past Volkswagens.

Many concept cars are shown without plans for production, but the powerful video presentation that followed conveyed a different message. Put together in brainstorming sessions among the Simi Valley group and Carolyn Lantz of SHR, the video showed evocative images of the old Beetle, such as a view of a surfer carrying his board, framed through the split rear window of a vintage Bug, and of other classic 1950s products.

On the soundtrack was soft violin music, then the words appeared:

It's funny the things we remember. The things we hang on to. The first day of school. A first dance. A first kiss. Our first car. Some things are simply unforgettable.

One little thing can bring it all rushing back. A song on the radio. The smell of suntan lotion. Seeing an old friend at the beach. The friend you could always depend on. Everything was a little less complicated then. Tennis shoes didn't cost $200. A jukebox played your favorite song. And a car was a part of the family. Right from the start.

The screen showed old Converse All Star sneakers and a beat-up baseball glove and the tonearm of an old record player bouncing hypnotically on a 45 rpm record. Images of the car appeared along with the words:

What if originality still meant something original? What if simplicity, honesty, and reliability came back again?

One look, and it all comes back. But then it never really left. The legend reborn. A friendship rekindled.

This was ad copy worthy of Roland Barthes, who turned the Citroën DS into a goddess, a cathedral, with a touch of Proust.

The car's keynote was the little bullet-shaped glass vase that fitted into the New Beetle's dash—the bud vase. With the vase, they had come upon the best symbol for the whole project: the rosebud of remembrance.

The logo was a tip-off to the essence of the car. Its three circles did with arcs what Reimspiess's original VW symbol did with angles. Little lapel pins with three intersecting arcs instantly became sought after items at the Detroit show, like the insignia of a secret society.

The pitch for the Concept One, complete with audio-video effects, context, and story, was also given to dealers—the people who have to love a car to sell it. A week later the company made the same carefully orchestrated presentation to some 400 dealers at a national convention in California. The dealers, eager for product fresher than the boxy economy cars they had tried in vain to sell to Americans for more than a decade, were overjoyed. Suddenly there was hope for their sad franchises. Each dealer was presented with a copy of Marsh's painting of the legendary birth of the New Beetle.

Within weeks dealers took the astonishingly premature step of erecting billboards showing paintings of the Concept One and the legend "future product." Auto billboards hadn't used paintings since the 1960s, dealers never announced cars they didn't sell yet, and "future product" was a falsehood since the company had not yet committed to producing a car based on the Concept One. The signs showed just how desperate the dealers were and how excited they were about tapping the public enthusiasm and selling a new Beetle.

Public response was equally enthusiastic. Within days of the Detroit unveiling, a woman sent a $500 check to Volkswagen as a deposit on a New Beetle. An executive framed and hung it in his office as a reminder of the public fervor.

Mays and Thomas were celebrated for almost single-handedly reviving a lethargic company. They understood the car, Americans said. They could see the German culture of VW better because they were outside it. But they were inside the culture of the *American* Beetle and it was this Beetle they were evoking in the Concept One. Only Americans would have exceeded their orders, formed their conspiracy—and heeded the call of the Bug instead of the boss.

At the Geneva show in March of 1994, the yellow Concept One sedan was joined by a bright red convertible, prompting speculation on whether

VW would build the car and what kind of car it would really be; in effect, how big, how expensive?

Even after the car became, in Mays's words, "a love child of the press and public," the response from Germany was deafening silence. Gradually, Freeman Thomas noted, the public pressure moved executives from negative to neutral, then to acceptance. Even Ulrich Seiffert began to admit that the car might be built after all. But the decision would be made at the top and as the months went on some feared that Piëch, distracted by problems in Germany, might continue the tradition of ignoring the American market. Finally word came from Wolfsburg that VW would build the car. The details were vague but the decision was clearly Ferdinand Piëch's.

chapter fifteen

he Germans were now confronted with the task of engineering a
car that could live up to the concept. It was an "emotional" car, they
said, and they were in the business of building "rational" cars,
rational in cost and shape and function. The idea of the New Beetle challenged the very basis of their professional lives. And they asked, Had
there ever been a car more rational than the original Beetle? Nobody in
Wolfsburg wanted to work on the new one. The car was such a political
hot potato that part of its development was farmed out to an independent contractor in Wolfsburg. For engineers, the good jobs, the ones that
would get you ahead, were on the new Golf—the car that had replaced
the Beetle as the core of the line.

The engineers proceeded to turn the emotional car into one that was
also rational.

As the engineers reported in, Piëch was confronted with a key decision. On which "platform" should the new car be built? Platform was
the most important word at Piëch's VW. To make the company profitable
again, he planned to cut costs by a third by building many cars on
common platforms—shared chassis and power trains. At a cost of $24
billion, Piëch aimed to create a line of fifty-five models on just five platforms. Different suspension and engines and setups could make these

cars drive in very different ways, but they shared many common parts from chassis members to water pumps and door handles.

The Simi Valley group had envisioned the New Beetle built on the platform of the Polo, an inexpensive car not sold in the United States, but after the first few experiments, Piëch made the critical decision: the new Beetle would be larger; it would be built on the company's most used platform, code-named the A4, which also served as the foundation for the VW Golf and Jetta, the Audi A3 and TT sports car, the Seat Toledo and the Škoda Octavia. This was the only way to justify producing a car essentially for a single market. No event in the car's history more convincingly demonstrated how much form prevailed over function in the New Beetle than the shift in platforms.

The other requirement for making such a small-volume car pay off— a car essentially for the American market, with planned sales of just 50,000, as compared with hundreds of thousands worldwide for the Golf—was to build it in Mexico, with its low labor rates.

In another cost-cutting move, the engineers put an older brake system on the front wheels, but in late August of 1995, Piëch drove a version of the car at the vast Ehra-Lessien test grounds north of Wolfsburg and was disappointed. He ordered the brakes upgraded, whatever the additional cost.

Not until the Tokyo automobile show in October 1995 did the company show a version of the car that was more mature than the Concept One. Called the Concept Two, this model was black, as if to appear serious and practical, and it had grown a foot to fit the new platform. Finally, at the Geneva auto show in March of 1996, a version of the car officially called New Beetle was presented and announced as the model that would enter production.

In the spring of 1994, still riding the enthusiastic reception of the Concept One, Freeman Thomas sketched a roadster of about the same size on a small scrap of paper. A few weeks later Ferdinand Piëch pulled the sketch out of a pile of concepts and ordered a full-size version built.

The Audi TT, as the car came to be called, after the Isle of Man TT race, was to the Concept One what the Porsche 356 was to the original Beetle: a sports car in spirit, with many of the parts of a people's car. Thomas was dispatched with a team of modelers and designers to Gamersheim, a small village near Audi's headquarters in Ingolstadt. There, in a secret studio, they worked around the clock to build the TT concept.

In the fall of 1995, at the Frankfurt auto show, the TT was presented, together with a metal and black leather press kit journalists called "S&M style." The shell-like TT was a carapace of a car that resembled a suit of knight's armor, flaunting the joints where its parts joined. Its big wheels gave it what Thomas called the look of a chariot and it was relentlessly mechanical in details, all cylinders and tubes. Inside, the brown leather seats were sewn whipstitch style like a baseball mitt—an American touch. It was inspired by Porsche 356 and Porsche RSK, and the Auto Union Type B racer of 1935.

The TT implicitly laid claim to the Porsche bloodlines—the heritage that Piëch symbolically brought with him when he was exiled from his grandfather's company.

The New Beetle had changed Volkswagen's image, but not its corporate culture. Within a year, both designers were exhausted and frustrated from their battle with VW executives. To Freeman Thomas it seemed that 10 percent of a designer's time went into design, 90 percent into selling the design. In 1994, Mays was promoted to the top design job at Audi on the strength of his success but left after just a few months. "I could already visualize all the cars for years ahead," he said. Thomas, also feeling the constraints of VW, left the company as well.

The pair established a new industrial design department for SHR, applying the ideas that had proved so successful with the Beetle to other kinds of products. They opened a California office a couple of miles from the Audi VW studio and went to work amid temporary furniture they had built themselves—desks made of doors laid over "Burro Brand" sawhorses.

"SHR stands for Saw Horses Rule," Mays joked. But could the New Beetle design really serve as a model for other products? How many companies could really change their whole personality by bringing back older shapes?

Clearly the New Beetle had radically changed the way cars would be conceived and designed. Ford soon hired Mays as its head of design and Chrysler hired Freeman Thomas to oversee future car design. The whole industry appreciated the creators of the New Beetle.

chapter sixteen

t he press loved the Concept One, and the repeated headline "the Beetle will return" suggested a car that was inexpensive and basic, but it was only the shape that returned. The New Beetle that evolved would be a car of limited practicality that cost almost as much as a full-sized family car—about $18,000 compared to $20,000 for the average American car. Company press releases began calling the new car an "upscale lifestyle vehicle." The New Beetle would be the old Beetle's ultimate doppelgänger. This was not an inexpensive, universal people's car but a pricey, stylish personal car.

The most winning feature of the Concept One was that small glass bud vase set into the dashboard. It was a symbol that blossomed into memory and association, it offered a neat evocation of the culture around the old Beetle—flower power. But by association, too, it blossomed into the full psychedelic rose of the Grateful Dead, whose fans had made the Bug and the bus their official mode of transportation.

The inclusion of the vase was also a fascinating piece of memory distortion and selective recall. The vase had not been a common accessory for the Bugs of the past. After World War II, an industry developed in Germany to supply accessories for the Beetle, as had happened with the Model T. The vases showed up in catalogues alongside such useful things as little hang-down pockets and ashtrays that slipped over the gearshift handles.

The originals of the flower vase were cut glass or porcelain. They had no bases and some clung to the dash like long, formless slugs; others were propped in rings. Some of the vases had flowers painted on them. They were silly, sentimental, romantic, and goofy at once, somewhere between gemütlich and kitschy. But their restoration in the new car lifted them above sentimentality to clever packaging.

The vase suggested the designer's original goal: to extract some primal automobile gestalt from the memory of the Beetle, to realize the child's sketch of a car. The result was as recognizable, as simple and strong as the circles of Mickey Mouse, the arches of Mickey D's, the image of a smile button or happy face. The car was like a toy, and one of Thomas's drawings literally depicted it as a wooden pull toy, as if assembled from semi-circular blocks. Thomas even drew the string with a little knob at its end, like the bright retro wooden toys popular with yuppie parents.

But were those three simple curves really the essence of the old shape? They gave the car a symmetry front and rear. "It looks the same coming and going," some said. That might have been just right for what the designers billed a car for the past and the future at once, a car imbued, as SHR's copy had it, with "memories of the future." It was a Janus car—looking forward and backward—a more educated observer noted.

But the original car was not symmetrical. Its dominant lines were the curve of the roof falling to the rear bumper and its echo in the similar fall of the hood—Hitler's "like a Beetle" line. The old car looked more like the profile of Alfred Hitchcock or the toy car Picasso sculpted into a monkey's head.

The key line of the old Beetle, its essential shape, was more like the line drawn in the old VW ad "How much longer can we hand you this line?" It looked like this:

It was as if a child were drawing the car or if you were drawing the car from memory. And that is what Mays and Thomas were doing: drawing it from memory. Even the colors of the new car were exaggerated the way memory exaggerates: the pastels of the original green and yellow were intentionally punched up into brighter greens and yellows in the New Beetle.

Despite their shared name and gestalt, in many ways the old and the new Beetles could not have been more opposite. One began from rigid functional specifications, literally laid down by a dictator, the other came from a shape, a logo in fact, into which the mechanics were forced. One had the engine in the rear, the other in front. One took its shape from function, the other its function from shape.

The shape of the original was rounded in a manner typical of advanced cars in the 1930s and 1940s. But its panels were stamped out with a pattern of raised lines on the hood and rear hatch that were decorative, almost baroque in their curves.

In the old car, the front hood was stamped to form two halves looking like the twin covers of a beetle's wings. The rear, with its vents and split, or "pretzel," window, was similarly stamped, a shape that added strength. On the rear panel, over the engine, the raised uvular shape and the divided window made the back of the car echo the pattern of the front. That double oval shape—the wings in front, the windows in back—implied motion, a shape running through the whole car. The windshield seemed to be drawn through the body into the rear window. Seen from above, the oval of the body played out again in the ovals of the fenders, but the body was also inflected by those stampings.

The original's interior was small but sheltering. Its walls felt rounded, as if puffing up its chest, and saying, I'm doing the best I can to make room for you.

It was a shape that echoed that of other cars designed by Ferdinand Porsche and Erwin Komenda, from the Silver Arrow racers to the postwar Porsche 356 sports car. Automotive historian L. J. K. Setright wrote of Porsche's cars:

> All of his vehicles have in common the same kind of curvature, the same appearance of thin fragile ultralight weight shells, enclosing large and superefficient voids. It was quite uncanny how Porsche could make something big, substantial and utterly open appear light and cavernous, as though it were a papier-mâché doppelgänger for the genuine heavy metal. Porsche's curves showed a sensitivity to the curves of the dishes and ocean liners and other objects around him.

The new car shared the same basic shape as the old only in the most general sense and the two represented different philosophies of design, production, and marketing. One was a universal product, aimed at mass sales and low prices. The other, its champions constantly emphasized, was not the reincarnation of the old Beetle but "a reinterpretation of an automotive legend" and an "upmarket lifestyle vehicle."

The new car had a water-cooled front engine and front-wheel drive. It possessed a near symmetry front and rear; the round red taillights and upturned line of the trunk lid replayed the round headlights and smiling hood line. Its shape was guided by an ideal of geometric purity. Kimberly Elam in *Geometry of Design: Studies in Proportion and Composition* asserted that the New Beetle's shape fit neatly inside a golden ellipse, the ideal Greek proportion of three to five, height to width, like the facade of the Parthenon.

Fitting an engine in front led to odd dimensions and strangely apportioned space. In front, the arched roof left a huge space above the driver's head, and much wasted space above the dash. A seven-footer would have been comfortable behind the wheel. (This, incidentally, represented a sort of ironic redemption of the famous Beetle ad of the 1970s, in which seven-foot-three-inch basketball star Wilt Chamberlain tried in vain to squeeze into the car. "They said it couldn't be done. It couldn't.")

In the rear, however, headroom was minimal and passengers were menaced by the rear window when the hatch was closed. Even children complained of being scrunched. The back seat felt even stingier to some people than the one in the original and it was smaller than the one in the contemporary Golf.

And the new car appeared thick and bulbous, very different from the fragile shell described by Setright. The full fenders, the deep cut into the bumper for the license plate, and the thick edges of the body extending past the rear deck lid all visually implied great thickness. The car looked puffed up, almost pneumatic. The front was shaped by curves instantly associated with a smile. "It makes people feel good about themselves. It makes people smile," the designers boasted.

This impression of thickness extended to the pillars supporting the roof that suggested strength and solidity. It was important for the new car to look safe, in contrast with the tin can qualities of the old Beetle. The appearance did not deceive: the pillars were part of a rugged structural safety cage that enabled the car to perform very well in government crash tests.

The arch of the roof was functional, as it rarely is in architecture anymore, and conveyed a primal sense of safety and comfort. Old Beetle owners had often remarked that the car made them feel protected despite its size and tin walls, as if they were inside a sphere. The exaggerated roof arch in the new car evoked that feeling more strongly. *Geborgenheit* was the untranslatable German word the designers in Wolfsburg used to describe the sheltering comfort of the arch.

The whole body seemed to be molded, not stamped like most cars. In most contemporary automobiles, the plastic bumpers are painted to resemble metal. But in the New Beetle, the waxy paint of the body and the size of the bumpers made the metal look like plastic.

And a lot of it was plastic—exotic new plastics inside and out, with chemical names such as polycarbonate/acrylonitrile-butadiene-styrene, polyamide, and diisocyanate. Their trade names—Desmodur, Cellasto,

Novodur, Lustran, Durethan, Kakrolon—suggested towns on some ancient map of Tartary. The Bayer Corporation boasted of "fourteen applications of engineering plastics, polyurethane foams and poly- urethane raw materials." Bayer plastics were used in the headlights and taillights, the auxiliary springs, and the interior. Inside the doors and dash were packed great bolsters of polyurethane foam. The new Bug was as padded as it looked.

We think of the twentieth century as one of the assembly line and mass production and the impersonal, standardized object. But it was also the century of a countervailing, compensatory effort to give character and personality to manufactured objects to humanize the products of the machine.

The distinction between water-cooled and air-cooled cars ran through the whole culture of VW buyers and buffs. The two have different maga- zines, different shows and festivals.

Air-cooled meant the classic, water-cooled meant a more conven- tional car. There was the implication that only air-cooled people were purists, that the air-cooled solution spoke of the simplicity and democ- racy of vehicles in a way that water cooling did not. In the Bug culture the distinction was as dramatic as the difference between warm- and cold- blooded animals.

We think of the twentieth century as one of the assembly line and mass production and the impersonal, standardized object. But it was also the century of a countervailing, compensatory effort to give character and personality to manufactured objects to humanize the products of the machine.

Characters to represent and sell things emerged in the curtain raiser to the century, the 1893 World's Fair in Chicago, where the Uneeda Biscuit boy in his rain slicker, Cracker Jack, and Aunt Jemima, the happy mammy who sold pancake mix, were introduced. Shortly after, the Michelin brothers were inspired to invent Bibendum, the tire man, as mascot. And in 1900 George Eastman adopted the elfin creatures of Hablot Browne as totems for his camera for Everyman—the Brownie. The Model T of pho- tography, the Brownie came on the market for a dollar, its box decorated with colorful lithographed Brownies holding images of the device inside—a device so easy a child could use it, we were told. "You push the button, we do the rest." Such products themselves became little characters. The abbreviations and stylizations used in toys and cartoons pointed to the basics of character in facial likeness.

Cars' personalities had often grown from their simple faces. The jeep seemed to grin through its grille; it took its name and persona from the knob-nosed creature in the comic strip *Popeye:* the army's GP (for general purpose) mutates into jeep, the general infantry's general-issue vehicle—the GIs' GP. In E. C. Segar's *Popeye* strip, Eugene the Jeep possessed magic powers to travel between dimensions. His nose and round bald head suggested Kilroy, the GIs' chalked calling card and mascot.

The Beetle's persona was like Kilroy, or Mickey or Charlie. In appearance, durability, and even driving characteristics, it resembled the character of the dogface soldier or the Little Tramp. "The great unlicked," as Walt Disney called Mickey Mouse, was also a good description of the car's performance. It showed persistence and surprising ruggedness but a lack of power and a tendency to be tossed about by the winds of change.

It is a human trait to project faces onto nature. We seek out our likeness; the abstracted face is a template through which we view the world. We impose it on living things and machines, on the leaf patterns of plants and the spots of insects, on parking meters and gumball machines.

Cars have faces by natural design. They are symmetrical, with two headlights for eyes and a grille and bumper for mouth. Occasionally someone adds a third headlight, like Preston Tucker or Hans Ledwinka in his Tatra, and sometimes the headlights become a bar or strip, and the change invariably jolts viewers as unnatural. The grille of the car is often designed as something like a mouth, and when the grille is omitted, many buyers miss it. They've gotten used to looking for a car's face. *Turn Signals Are the Facial Expressions of Automobiles* was the title of a book by Donald Norman, but even at rest, some cars frown, some squint with power and determination, some keep a stiff upper lip of noncommittal efficiency. And some smile. The New Beetle actually had two faces: the one in the rear, with its round taillights and trunk line, seemed to smile just as the front did.

Stephen Jay Gould in a famous essay on Mickey Mouse showed that humans respond to cartoon creatures—and to living animals—according to another template of the face. He cited the work of Jan Tinbergen and Konrad Lorenz, which demonstrated that human beings had something like a hardwired response to cute animals. The large crania, wide cheeks, and big eyes of certain animals mirrored the proportions of a human baby's face—and inspired nurturing. So Mickey, similarly proportioned, is cute and sympathetic. We want to take care of him. He makes us smile.

Gould writes that "the abstract features of human childhood elicit powerful emotional responses in us, even when they occur in other animals. . . . When we see a living creature with babyish features, we feel an automatic surge of disarming tenderness." Mickey, he noted, "devolved" from a more mischievous, more adult character in the 1930s to his cuddly later self. He was cuter in the 1950s than he had been in the 1930s.

Mickey's key shapes were "a relatively large head, predominance of the brain capsule, large and low lying eyes, bulging cheek region, short and thick extremities, a springy elastic consistency, and clumsy movements."

These shapes offered behavioral "cues"—the same word car designers use to describe brand traits. Biologists call them "innate releasing mechanisms." Konrad Lorenz experimented with substituting objects with "faces" for other creatures and found the same pattern. "Even inanimate objects that mimic human features can have powerful effects," he concluded. "The most amazing objects can acquire remarkable, highly specific emotional values by the 'experiential attachment' of human properties." Machines or other man-made objects can elicit friendliness or hostility. Even landscapes can evoke expressions, as in the menace read into cliff faces or storm clouds.

With its big head—high arched roof, low eyes, pudgy fenders—the New Beetle perfectly fits the formula for the lovable creature. And its relation to the leaner, less cute original is similar to that of the cute 1950s Mickey to the 1930s Mickey. What they had in common was the smile.

chapter seventeen

Volkswagen introduced the New Beetle to the press in Atlanta in February 1998 with a 1960s theme party. It was a celebration of branding amidst a cartoon cavalcade of the era: huge pictures of John Lennon and Muhammad Ali, Jimi Hendrix and Janis Joplin, rebels now rendered flat and half-toned. Most of the guests donned tie-dyed T-shirts. Some had their faces painted by artists hired for the occasion. A Janis Joplin impersonator sang about her friends all driving Porsches, but changed the lyrics so they torturously emerged as, "Oh, Lord, won't you buy me a New Beet-le."

In the middle of the room, the chief operating officer and director of the board of the Volkswagen Group, the onetime master of Porsche's racing team, the man credited with developing the redoubtable six-cylinder motor in the Porsche 911, the pioneer of Audi Quattro four-wheel drive and the W-12 super engine, stood in tie-dyed clothes and love beads. But the peace sign Ferdinand Piëch wore hung upside down so that it resembled a Mercedes tristar.

A faint smile played about his lips, as if he knew something but wasn't telling, and he said very little.

He stood beside one of the New Beetles, and the journalists asked him about the small back seat. "I don't think even kids would fit in there," one said.

"I have three kids," he said, staring down the questioner. "They fit."

Faced with a father as intimidating as Dr. Piëch, what child would not fit?

After five years as chief of Volkswagen, Piëch had a lot more than the New Beetle to boast of. In Atlanta, he could already point proudly to a turnaround at Volkswagen. When he took over, the company was losing $2 billion a year and VW's overstaffed plants could not be run profitably even at full production. He came to an accommodation with the unions, cutting back hours and establishing a four-day workweek in return for preserving jobs.

Using his platform strategy, Piëch had cut costs, trimmed payroll, and improved quality. He instituted just-in-time methods, first at the factories of Seat, its Spanish arm, and then at the VW factory in Puebla, Mexico. The bread-and-butter Golf fell from thirty-two man-hours to twenty in production time. In just two years the two-billion-dollar loss of 1992 turned into a profit and quality ratings improved fourfold.

He expanded the Volkswagen Group—creating a "Greater Volkswagen"—by purchasing a number of foreign auto companies. To VW, Audi, and Seat, purchased in 1986, Carl Hahn in 1991 had added Škoda, once famed for its cannons and tanks, in the Czech Republic. Piëch went further. He expanded the Chinese operation, a joint venture with a local company; soon it produced 400,000 cars a year, half of China's total output. He purchased Lamborghini, the Italian sports car company, and when it became available in 1998 he bought the rights to the name Bugatti. In a major face-off with BMW he bought Bentley, a marque beloved of anglophile Germans. It was Piëch's favorite car; soon he was driving a blue one from his home in Braunschweig to his office in Wolfsburg.

The platform strategy meant the company could share parts—only four clutches for all the engines in all the cars, only a few kinds of steering wheels. These economies of scale gave the company leverage in negotiating with parts suppliers. VW was placing larger orders and Piëch demanded price cuts for volume.

He ordered his engineers to tear down, say, a Bentley and examine it part by part. He made sure that the same screws and bolts a supplier was charging Bentley a premium for—after all it was a Bentley!—would now be purchased at the much lower price VW got. "You won't believe how little I pay for a steering wheel," he once bragged.

He also added value to the lower level cars. Silicon-damped handles that slipped back softly and pleasantly had previously been a luxury feature; with purchasing deals that lowered the unit price from $5 or so to 75 cents, they were installed on the less expensive VWs as well.

Piëch was a stickler for detail and a hands-on chief engineer. When the updated Passat was introduced, the first car developed during his regime and the vital family-sized model, he drove each of the sixty vehicles that were to be test-driven by the press to make sure each was in good shape.*

But Piëch's tenure was consumed by a controversy. One of his first acts in March 1993 was to hire José Ignacio López de Arriortua, the parts wizard from General Motors, who claimed to have wrung some $7 billion in costs from GM's budget by striking harder bargains with suppliers.

López was accused of bringing with him to VW confidential GM information on how to build thousands of specific parts in the least expensive way possible, plans intended for Opel's new O car. An aide of López was caught shredding GM documents; German and American authorities began assembling criminal cases against López, and General Motors launched civil suits in the U.S. and Europe. For a time in the summer of 1993 it looked as if not only López but Piëch himself might lose his job.

López had long cherished the ambition of bringing a modern automotive plant to his home, the Basque area of Spain. He seems to have been playing GM and VW off against each other to obtain such a plant, and VW, as the owner of Seat of Spain, seemed the better bet to deliver.

López imagined a new type of auto plant where many subcontractors would build sections or modules of a car in buildings adjacent to the main assembly plant. This "assembly park" would lower costs. VW built a factory to apply this system, which López nicknamed Plant X, in

---

* This time Piëch made sure the Passat fit the American market—it had a zippy turbocharged four-cylinder or a full-sized six-cylinder engine, electric windows and other gadgets, and something Germans did not understand or desire—cupholders.

Resende, Brazil. López's ideas were also instituted in VW's plant in Puebla, Mexico.

The cases against López and VW dragged on until the fall of 1996 when he resigned. In January 1997, VW agreed to pay GM $100 million and made a lukewarm apology to settle the civil actions. López never got his Basque plant. In 1998, he was involved in a serious automobile accident and spent three months in a coma. Eight years after López left GM, American prosecutors were still trying to bring him to trial.

The controversy made Ferdinand Piëch, never at ease with the press or public, more hostile and reclusive. Those who rose in the company were the men who helped him during the controversy, lawyer Robert Buchelhofer, and public relations aides Klaus Kocks and Otto Wachs. Piëch's dropping in at the New Beetle party was one of his rare public appearances.

His presence, too, was a reminder that the Bug's story was a family story.

The Porsche and Piëch families had created VW and Porsche, which was always something like VW's cousin. Ferdinand Piëch continued to hold a huge chunk of Porsche stock even while working at Audi and VW. The two companies often worked together over the decades, and sometimes jointly developed vehicles, such as the Hans Gugelot–designed 914 of the early 1970s or the sport utility vehicles planned in the 1990s, the Volkswagen Colorado and the Porsche Cayenne.

But there was a deep rivalry between Ferdinand Piëch and his cousin Ferdinand Porsche. Butzi and Burli—the two men and the caricatures that emerged in auto industry gossip about their characters—exemplified two different approaches and philosophies. Ferdinand, or F. A. "Butzi," Porsche meant design; "Burli"—Ferdinand Piëch's nickname—stood for engineering.

Butzi would forever be associated with the lovely classic body shape of the 911. But Burli would be credited with the development of the engine that made it go, the air-cooled six-cylinder boxer, a masterpiece like the Beetle's venerable motor.

The conflict between the two men was tied to the conflict between the families. Piëch never believed that Ferry Porsche, Butzi's father, son of the founder, had given him proper credit for the 911. While those inside Porsche liked and admired Butzi, they did not believe he had the character of a corporate leader. Piëch, on the other hand, was admired for his forceful ways and his engineering skills but not liked.

Both had jockeyed to succeed Ferry Porsche as head of the company. In 1972 the conflict between the two men nearly killed Porsche. When the company's board banned family members from its top job, F. A. Porsche left to establish Porsche Design, an industrial design firm. He took with him a special right to use the Porsche name on accessories like sunglasses and key chains and he also designed other products—Grundig radios and Samsung cameras. In the 1980s Steve Jobs even tried to hire Porsche Design to rework Apple computers.

When the New Beetle arrived, F. A. Porsche was designing a Porsche coffeemaker and a Porsche toaster for Siemens, the German conglomerate that would also produce parts for the New Beetle.

At first blush, the New Beetle's extroverted personality could not have seemed more different from the reserved Ferdinand Piëch. But by building that car and restoring Volkswagen's fortunes, Piëch was implicitly showing himself the true heir to his namesake and grandfather, Ferdinand Porsche.

With the press Piëch appeared uncomfortable. His eyes would brighten but he came across as haughty. He seemed to be keeping a secret. But if his cutbacks in employment and rationalistic reduction of basic car platforms seemed the work of a coldhearted engineer, he was a visionary when it came to more efficient vehicles. He furthered development of the direct-injection turbodiesel engine and championed the little Lupo car, which got about 92 miles per gallon using it.

By 2000 shrewd industry observers wondered whether he had not gone too far in reducing the number of platforms, whether he was not blurring lines between brands when, for instance, he planned a luxury Volkswagen that seemed to compete with Audi, the luxury brand of the greater VW family. The move recalled the mistakes General Motors made in the 1960s, when it created luxury Chevrolets that stole the thunder from Buick and Cadillac. Many analysts believed that Piëch's arrogance and ambition would inevitably lead him to overreach.

The day after the New Beetle party in Atlanta, the press cars were arrayed in military formations in front of an Olympic symbol at the former equestrian grounds for the 1996 Olympics. The joined rings in combination with the ubiquitous VW and New Beetle logos could not help but recall Leni Riefenstahl's film *Olympia*.

The party was an orgy of brand obsession. VW was increasingly wrapping itself in the marketing and the cult of the brand. Buzzwords from branding consultants had come to occupy the place in marketing thinking in the 1990s that those about psychology held in the 1950s.

The New Beetle, Piëch declared in his speech, "cannot deny its origins and the magic of its shape," he said. It was a curiously romantic sentiment.

Some heard him to say instead "in the magic of its shape," which would have been more accurate. "It was here that the car first became known by the nickname Beetle," he added, forgetting the history of the 1930s. "It should look like a beetle."

"It gives people joy," he said of the car. Joy through strength? Joy through design? Joy through brand strategy?

Piëch delivered the sentences tentatively, as if he were still getting used to the idea.

He and other VW executives were just grasping the whole American notion of brand—and being seduced by it. Soon it came to obsess them.

In Atlanta a whole range of products bearing the New Beetle logo were released: clothing, key chains, necklaces, flashlights, watches, even a boom box radio with a bud vase attached. Best of all were the sunglasses in the shape of the New Beetle logo. The two small arcs formed the top of the lenses, the larger arc joined them. It was a perfect metaphor for Volkswagen's new view of the world: looking at it through the logo.

The company chronology included in the press kits that day managed to avoid any but the briefest mention of the 1930s. Stranger still, the Atlanta party evoked a caricature of the American 1960s, just as the New Beetle caricatured the American memory of the old Beetle. The first time as tragedy, the second time as farce—Marx's famous phrase was inescapable. Or as one cynic at the party said, "If you can remember the 1960s you weren't really part of it."

The Germans had not been part of this version of the decade. Their 1960s was the decade of the Cold War, the Berlin Crisis, and the Baader-Meinhof terrorist gang. The Beetle was a relic of painful times, happily left behind, and they would be cool on the New Beetle when it arrived in their showrooms.

chapter eighteen

I n America the New Beetle was a success long before it ever arrived at dealers. Its image had been reproduced many times between the showing of the Concept One in 1994 and the time it actually hit the market in 1998. That year, the company sold 55,842 New Beetles—more than the hoped-for 50,000.

The New Beetle was also pulling people into the VW dealerships for the first time in years and helping Volkswagen sell some 25,000 more of its other models than the year before. It was working as a "halo" car, with "rub-off" on other cars and the whole brand, "a silver bullet product" in the phrase of Barry Shepherd of SHR, the firm that had helped Mays and Thomas position and promote the Concept One.

The success could not have come without other factors: new ads, improved quality, better management from Clive Warrilow, who took charge of VW in the U.S. in January 1994, the same month the Concept One was shown, and the fact that top executives such as Jens Neumann urged Piëch to fight for the U.S. market rather than abandon it. But more than anything else, the design of the New Beetle changed the way people thought about Volkswagen.

The first signs of the rub-off effect came in the fall of 1997 when the updated Passats began to appear in VW dealerships. The company proclaimed a family resemblance between the Passat and the New Beetle. The arched roofs of both cars offered a feel of being sheltered, their promotions noted, expressed in the German word *geborgenheit*. But the New Beetle was to signal something had changed about VW in general: "There's a bit of Beetle in every Volkswagen," was the appeal.

The New Beetle also helped to sell the Jetta, its near twin under the skin. The Jetta, a design curator in San Francisco commented in 1999, had become the car of choice for young Silicon Valley professionals.

Although the characteristic colors of the New Beetle were not the muted pastels common in the old Beetle, buyers were not taking to the bright toy red and yellow models. To the surprise of the company the hit was the racing silver that harked back to the Auto Union and Mercedes racers of the 1930s. Silver implied that the car was a piece of equipment, not a toy after all.

It was said that with the Volkswagen in the 1950s the Germans sold an American idea—Henry Ford's idea—back to the Americans. And with the New Beetle, in the 1990s, the Americans sold a German idea back to the Germans.

But the Germans weren't buying. When the car came on the market in the fall of 1998, despite another 1960s theme party for the press, which cost nearly $3.5 million, sales were slow. Compared to the practical Golf, the New Beetle was an expensive novelty item. It was priced at about $22,000 in Germany compared to $17,000 in the U.S. The Wolfsburg executives had always anticipated this response and introduced such special models as the "en vogue" New Beetle, with richer interior leather and fashion colors. The strategy confirmed that the market for the New Beetle in Europe was small but fervid.

In the U.S. the New Beetle also soon acquired what James Hall, the noted automotive analyst, called "third order publicity": its success was marked by its role as the top sweepstakes prize in dozens of contests for fast food chains and other enterprises. It showed up in the background of advertisements and fashion images as a token of hipness and fashion.

Its shape and name were licensed by toy makers. It figured in driving video and computer games. There was soon a Barbie Beetle and a radio-controlled version, like the cars the New Beetle's designers had played with in the parking lot behind the Simi Valley studio.

The VW ad campaign from the Boston firm of Arnold Communications had begun to change VW's image even before the New Beetle arrived. The "Drivers Wanted" ad campaign created by Steve Wilhite and Liz Vanzura of VW and Ron Lawner and Lance Jensen of Arnold premiered in the summer of 1995. It was aimed at buyers in their twenties and thirties. The slogan appeared to exclude while subtly including: Who would admit to being a mere passenger "on the road of life"? "Drivers" was a synonym for active people and flattered buyers who liked to think of themselves as better-than-average drivers. The ads aimed at the minority of customers who bought aggressively, not just by consulting the *Consumer Reports* charts or J. D. Power surveys of defects per hundred cars but according to style and sensation. "German engineering" and "European handling" implicitly pitched VW as the un-Japanese car and aimed the ads at customers who did not want Corollas and Civics, the safe reliable choices sold in huge numbers but with little character or soul.

Joint marketing campaigns were developed with K2 skis and Trek bicycles, popular with the customers VW was aiming for. Special K2 and Trek edition Golfs came with skis and bikes. The critical ad in the "Drivers Wanted" campaign came well before the New Beetle arrived. The so-called "Sunday Afternoon" spot of 1997, better known as the "Da Da Da" ad after its music, showed a pair of slackers cruising around in a Golf looking for old furniture. Accompanied by the crudest melody possible, a catchy snatch of song grabbed from an obscure band called Trio, the pair stop and load up a discarded easy chair. They drive around a bit, then wrinkle their noses and look at each other, finally dumping the smelly piece of furniture.

In its ads for the New Beetle Arnold borrowed inspiration from DDB's old Beetle ads the way the New Beetle took off from the old. Both television and print ads employed the generous "white space" of the old print ads; many of the shots were almost empty, with just a few words and tiny images.

The slogans practically wrote themselves. "More power, less flower." "0 to 60—yes." "If you sold your soul in the Eighties, here's a chance to buy it back." "Reverse engineered from flying saucers."

"The simplicity of the design begs you to get out of the way," declared Ron Lawner, Arnold's chief creative officer, and that was mostly what the agency did. They deferred to Mays and Thomas's design. "How could they miss?" asked ad critic Bob Garfield in a review headlined "Beetle Could Cruise Even Without Fine Ads."

The advertisement that revealed the most about Arnold's strategy, however, was one run by the advertising publication *Media Week*. "This is not a car," read the caption beneath an image of the New Beetle. The copy explained that the car was not a car because it was an idea, a token of the brand. Lawner was quoted describing the buyers of the New Beetle as "belonging to a club." Owning the car meant "wrapping oneself in the brand," he crowed. "It is an extension of your personality."

Arnold and VW were returning to the earlier ad fad of "affinity marketing." VW's marketing department created a real club and solicited New Beetle buyers to sign up for its newsletter and promotions. They established a Web site, to which thousands contributed decorating suggestions for the New Beetle—paint it like an Easter egg! a soccer ball! a rolling version of the *Mona Lisa*! The photographer Helmut Newton, known for his erotic fashion photographs, shot a series in which the New Beetle figured as a lusty prop. The company, which had long held itself aloof from the wide but weird VW owner culture, began to sponsor car shows and contests of customizations. Arnold and VW's marketing bosses, Steve Wilhite and Liz Vanzura, were trying to do by design for the New Beetle what the old Beetle had done for itself—confer a special status on its owner, a quality of being in the know. The ads were funny, subtle, full of actors who looked like real people, not models. They were about a sort of geek chic, but an implied chic nonetheless.

The first New Beetle ads seemed aimed at nostalgic boomers who recalled the original car, and the first buyers included mostly people in their forties and fifties. But soon the New Beetle acquired a set of buyers and buffs who were younger and for whom the old Beetle was simply a distant echo. These younger buyers loved the pod-on-wheels quality of the car; they would have bought the New Beetle even if the old Beetle had never existed.

By 2000 sales of the New Beetle had leveled off. In 1998, VW sold 55,842 New Beetles and in 1999, the first full year of production, 83,434. But the number fell to 81,134 for 2000 and in 2001 all the way to 60,891. Had it succeeded on the strength of a chubby-cheeked cuteness that buyers were growing tired of?

The New Bug's very fashionableness threatened its longevity in the market. The old car, after all, had been the antithesis of fashion. The company had expected there would be a flurry of initial popularity; the challenge was to maintain sales over the long term by constant freshening—"Life cycle management" the marketing mavens called it. New colors, new features, whole new models were introduced.

To keep the New Beetle hot, Volkswagen borrowed tricks from Swatch and other fashion companies, offering "limited edition" colors with names like Vapor Blue, Reflex Yellow, Isotope Green, and Snap Orange.

The company noted that 60 percent of the cars were purchased by women, who responded to the Bug's endearing quality. To appeal to men, VW tried to give some versions of the car a more rugged personality with the addition of aerodynamic panels, new wheels, and larger tires. It set up a racing series and introduced a more powerful, 150 horsepower turbocharged model distinguished by a rear wing, or spoiler. In Europe, VW issued the RSI model with a powerful 225 horsepower VR-6 engine, produced in an edition of just 250 a year for racing. (The RSI was never brought to the U.S. because fitting the larger engine kept the car from meeting U.S. front-end impact standards.) In 2002 came the Turbo Sport model, with a front modified to look something like a Porsche 911, extra horsepower, and a computer-controlled electronic stability program. VW played another change in the concept car called the Dune, a dune buggy variant produced by VW's German design staff. These tough, macho models evoked a Bad Bug, like the villainous anti-Beetle, the Darth Vader on wheels that showed up in Disney's 1997 television remake of *The Love Bug*. But the convertible, the best bet for rejuvenating interest in the New Beetle, was delayed by technical problems and would only arrive late in 2002.

The New Beetle naturally increased interest in the old Beetle. Firms selling spare and substitute parts for the old car increased their sales and added items for the New Beetle. Local VW clubs, which had flourished since the late 1960s, and their "Bug-ins" and "Bug-outs" and other gatherings grew more popular.

Some fans of the Beetle wanted more of it brought back than they found in the New Beetle: in Munich, two brothers named Rosenau imported old Beetles from the plant in Mexico, upgraded the engines and emissions systems to European standards, and sold them for $20,000 to $60,000. (An original Beetle sold for about $7,000 in Mexico.) A firm in Arizona began to do the same thing for the U.S. market.

Prices of used old Beetles did not significantly increase, but there was new interest in older and rarer models—pre-1953 split window models, Things, even cars from the Nazi years. A few collectors and buffs, however, revel in these despite or because of their Nazi associations, often restoring KdF-Wagen when they can be found and Kübelwagen, made authentic right down to the paint-stenciled palm tree and SS logo of the Afrika Korps model.

One of the surviving cars, astonishingly, was manufactured on March 5, 1943, and delivered on March 20 to Hitler's Chancellery in Berlin. It's not clear who owned it; perhaps it was attached to Hitler's motor pool. It eventually fell into the hands of Eberhard Lerbs, an East German collector who rented vehicles to filmmakers. In 1991, when the aging Lerbs dispersed his collection, the car ended up in the hands of an Arizona collector.

It was driven from East Berlin to Hamburg, shipped by container to Los Angeles, and trailered to Tempe, Arizona, where it was restored. An article in *Hot VWs and Dune Buggies* magazine trumpeted the specimen's "ultra-rare Bakelite horn" and noted that the "aluminum spring plate covers are very desirable KdF issue, complete with cogwheel logo. Even VW's Museum car doesn't have these!" The German supplier was found for salt-and-pepper upholstery fabrics "but what may be the rarest part on the car is the VW crest, which is DAF issue," and is "an item that many enthusiasts drool over."

The car was driven some 5,000 miles as part of a collectors' "Return to the Fatherland" tour in 1999 and shipped back across the Atlantic and driven to Wolfsburg and the site of its manufacture.

It had taken the old car many years to become the subject of customization but almost overnight the New Beetle attracted owners who changed its appearance. Companies found the car's catchy shape a handy form of rolling billboard and purchased Beetles to decorate with bright colors and flashy graphics. A Florida seafood restaurant dropped a huge plastic lobster on the top of a New Beetle to advertise itself. The Tropicana orange juice company painted one car up as a huge orange to promote its products. Other companies painted up New Beetles to look like insects or frogs. Soon individual owners followed.

One day in August 2000 at a New Jersey parking lot, a midday mist hung over acres of Volkswagens and merged with its auditory equivalent: throbbing bass, the generic blend of deep tones in the music seeping out of the cars. There were Volkswagens of all kinds: original Beetles and ancient buses, notchbacks and Sciroccos, Corrados and Golfs, Jettas and GTIs.

Old Beetles sat together in resplendent restoration, all in muted tones of blue and green and cream, while the shrill colors—electric green and dazzling yellow—and louder music of younger people demarcated the New Beetle quarter.

Festivals such as this took place almost every weekend across America and around the world. They were called Bug-ins and Bug-outs, festivals and funfests, and madnesses. Some were famous among the buffs: the Sacramento Bug-o-Rama and the Shangri-la in Prescott, Arizona.

There were stands selling accessories, nipple hubcaps and EMPI-style spokes, bumper brackets and gravel guard aprons, Cal Look mirrors and

Monza-style exhaust pipes and Tee Pee Exhausts. Old parts were offered, rusty turtlebacks of hoods or shells of whole cars, stacked up in pieces like a post-crash autopsy, worn badges and chrome strips and whole new engines. On one seller's table sat an array of old-style, porcelain flower vases for the dashboard. Another man sold T-shirts printed with the old DAF cogwheel and swastika logo.

The company's marketing department hoped to foster the same clublike affection for the New Beetle that the old one had long enjoyed. VW's marketing team sponsored gatherings at beaches in California and New York with rock and roll bands and handed out awards to the best-customized cars. To some longtime Bug buffs, aware of the company's long policy of holding aloof from the clubs, these seemed artificial. More interesting were the gatherings the New Beetle inspired on its own. A group of New Beetle owners got together in Roswell, New Mexico, site of the alleged 1947 UFO crash. Their gathering was built on the conceit of the slogan in a New Beetle ad: "reverse engineered from flying saucers." The marketing men, as usual, were always one step behind the designers and the owners.

At all these Bug fests, the Bug was mutating as owners tried to give the cars personality—their personality.

New Beetle owners turned their cars into Love Bugs and Lady Bugs with feelerlike antennas, soccer ball Bugs and baseball Bugs with seams like a ball. There was a green Beetle on whose dashboard sat a half dozen models in the same green, each half the size of the one beside it, a Swiftian recession in scale, the shape subtly growing rounder and, well, buglike, with each diminution. There was the "Bug's Life" model, inspired by the film, the M&M model, with plush toys of the round candy. There was the inevitable Herbie model, painted with red and white and blue stripes and number 53 just like in the movies. It belonged to a woman from Delaware who regularly traveled up and down the East Coast with her car.

Nearby stood another car painted with the number 52. The car belonged to a man named Ted Mendez, aka Ted Smooth, a disc jockey who lived in the Bronx. The 52 stood for May 2, his birthday. On the car's front was a strange, airbrushed image, a child's face—Ted at the age of three, his inner child seeming to burst through the sheet metal. He looked old and serious and a little bit like the actor Peter Lorre. It was the picture from his passport, which his father, an auto mechanic who abandoned the family, had signed before the family came to the United States from the Dominican Republic. The same image was tattooed on Teddy's shoulder. Under the photo was the inscription "est. 1971," the year of his birth. Ted called the car "my silver son."

Driving past a VW dealership in Queens, he had seen the bright silver Bug out front and immediately fell in love with it. When he got home he phoned the dealer and bought it—the display model, right off the stand.

On the spot Teddy had formed an image of how he wanted to make the car look. He added body panels and a rear wing and new lights and redid the interior with red and green leather seats, a homage to the colors of the TAG Heuer luxury watch company. He put in a powerful sound system and video games.

When he finished, the car was too valuable to drive except to the shows where he and his support crew of friends set the car up with a display of its awards.

Teddy Mendez had only the vaguest memory of the old Beetle. It was a thing from the past and, like the TAG Heuer logo, a shape to sample, as he sampled bits of music in his work as a disc jockey.

That the New Beetle elicited this level of affection fulfilled the wildest dreams of its creators. It had taken on a life of its own.

The press was filled with glib analysis of the New Beetle's success in which the car was tagged as "retro," a word J Mays and Freeman Thomas loathed as an oversimplification.

A March 1998 cover story in *Business Week* on "the nostalgia boom" compared the New Beetle with the retro-style baseball parks, Camden Yard in Baltimore and Jacobs Field in Cleveland, and designed communities such as the Disney Company's Celebration, Florida.

Retro was tied to the revival of old TV shows and the arrival of the Restoration Hardware stores, which brought back toys like the Slinky and wooden Duncan yo-yo of 1954. Retro, pop sociology had it, was the aging baby boomers' attempt to retrieve the idealized relics of their childhood and teenage years. Retro sought to return to a simpler, more optimistic time.

Other retro cars appeared. Renault experimented with a retro sports version of the 4CV called the Fiftie. There was inevitable speculation on a "New 2CV"—the classic French people's car rethought. In the U.K., MG came out with a retro sports car. The British Mini was reborn, New Beetle style, without the original ever going out of production. The retro version, which arrived on the market in 2001, took a somewhat different approach from the New Beetle. The original, introduced in 1959, had been in production for forty years, rivaling the Beetle in popularity and compact, clever efficiency. Like the Beetle, it became an object of affection because it was basic and unchanging. The successor, like the New Beetle, was aimed at a select high-end market but it was designed on the premise: what would the car have looked like if it had evolved?

But part of the appeal of the Mini was that like the original Beetle, it changed little and grew to be an object of affection because it was basic and unchanging.

The original, the masterpiece of designer Alex Issigonis, was created in response to the 1956 Suez crisis and resultant fuel shortages in the U.K. It firmly established the traverse-mounted front engine as the conventional arrangement for future small European and Japanese models. It had more interior room than a Beetle, but was easier to park in European streets. Like the Beetle, the Mini carried counterculture associations. It was the miniskirt of cars. Ringo Starr modified a Mini to carry his drums. Actor Peter Sellers owned several. The Mini didn't have a Love Bug film, but it did have the crime caper *The Italian Job,* with Michael Caine, in which the thieves escape in Minis.

The new Mini struck the same chords of cuteness and nostalgia, mixed with reassuringly up-to-date technical and safety features. The designer, Frank Stephenson, drew imagined 1970, 1980, and 1990 models before finally creating a real 2000 one.

Chrysler took the retro concept further with the PT Cruiser. While other retro cars, like the Beetle and the Thunderbird that J Mays designed after moving to Ford in 1997, took inspiration mainly from a single model, the PT Cruiser's historical associations were more free-floating, even "primal."

It was part gangster car of the 1930s, part woody wagon, part delivery truck, with a nose borrowed from the Prowler retro hot rod. This was a more generalized retro, guided by the psychologist Clothaire Rapaille, who wanted to "push the reptilian hot buttons" of drivers.

Rapaille frequently used the term "imprinting." Growing up in wartime France, he recalled, he had imprinted on the GIs who arrived after D-Day in Jeeps, bearing Hershey bars, and came to love American culture. He began his career working with autistic children, then shifted to consulting for Kraft, Coca-Cola, and Kellogg's. His eclectic references included Freud and Rudolf Arnheim, but above all Jung. He intuited that the essence of Folger's coffee was aroma. He delved into the secret meanings of barbecue sauce and wrote about the archetype of cheese.

Rapaille interviewed typical buyers while they lay on floor mats half dozing and free-associating. The PT's shapes, he believed, reached down to some subliminal layer of automotive evolution deep in the brain stem. At least they managed to evoke movie and TV images—*The Untouchables*, or *Beach Blanket Bingo*. His approach impressed David Bostwich, a marketing manager at Chrysler, and Bob Lutz of Chrysler. Dr. Rapaille's sessions revealed that people felt they lived in a hostile world. The PT Cruiser's full fenders implied protection. The upright windshield and trucklike back combined archetypal film-derived visions of surfers and gangsters—for weren't American archetypes lent by Hollywood?

The PT Cruiser was built on the platform of the Neon subcompact, which Chrysler launched in 1994 in high hopes that its round headlights and cute face would inspire the sort of affection the old Beetle had. It did not, but suddenly the PT Cruiser became the next "it" car, as J Mays put it, given away as a sweepstakes prize and featured in fashion shoots the way the New Beetle was. Chrysler sold 150,000 PT Cruisers in 2000, its first year.

Just as the Third World seemed about to adopt the Fordist idea, decades after the First and Second Worlds, the universal automobility that Europe and America took for granted came under attack.* Automobile ownership had soared to proportions unthinkable even a few decades ago. In some states of the United States, there were more cars than licensed drivers. More cars meant more traffic and more air pollution.

The New Beetle was a success not only in the marketplace but as a landmark of design that changed the way carmakers and other manufacturers thought about their products. But neither the New Beetle nor the cars it inspired brought the same technological or social innovations of the old Beetle. It was not the technologically revolutionary car it could have been. Where were the engineering ideas behind the Concept One? What happened to the innovative, "green" alternative motors that had given "lightning bug" its earliest justification in the Simi Valley studio? Was it possible to produce a new people's car that would keep alive the ideal of the personal car for ever greater masses without destroying the atmosphere or paving the earth in the process?

In the last years of the century with an eye to the expanding markets of the Third World, some manufacturers began to think about such a car.

In India, where automobile sales were predicted to rise fortyfold by 2020, Ford offered the Ikon, an efficient small car with a higher roof to accommodate native headgear and an engine resistant to monsoon rains.

China, where sales were expected to rise ninety times during the first decades of the century, sought proposals from Western manufacturers. Porsche offered a model in the mid-1990s, and in 1997, Chrysler offered a modern-day version of the Deux Chevaux, which, according to a Chrysler executive, François Castaing, would be "as easy to assemble as a toy."

The prototype, developed from what was called the Composite Concept Vehicle designed by Bryan Nesbitt, who would design the PT Cruiser, was made of two composite plastic shells attached to a metal frame with

---

* When the film of *The Grapes of Wrath* was shown in Europe in the early 1940s, local audiences were baffled by the images of the Okies heading west in their battered jalopies. What was the problem? the European viewers asked. How could these people be poor? They even owned automobiles.

a handful of bolts. The shells were made of recycled plastic bottles, the color molded into the plastic to save the cost and environmental impact of spray painting. Weighing only 1,200 pounds, powered by a simple two-cylinder engine, it would run no faster than 50 miles per hour. It was designed for the Chinese countryside—with a high clearance for rough roads, easily maintained parts, and a body that could be washed down, inside and out.

But the Chinese officials wanted something different. They finally worked out an agreement with General Motors for a much more conventional car to be made in China. They insisted the car be branded not General Motors or Chevrolet, but Buick. It was not the name of a people's car but that of a wealthy man's car and it held magic for them.

Perhaps small could also become beautiful in new ways in the developing world. An energy expert and disciple of E. F. Schumacher named Amory Lovins preached a theory of what he called the soft energy path, one side of which would evolve into a vision of a new kind of people's car, a set of specifications as revolutionary as those of the Bug. Lovins called these cars "hyper-cars." They would have engines that were hybrids of diesel and electric, bodies of light composites, regenerative braking, and sleek aerodynamics. They would get fifty miles per gallon or more.

In Europe, in 1993 Renault introduced the novel Twingo, which reduced the shape of an American minivan to the length of a compact, gave it a frog face of upright headlights, and captured the hearts of young French families. It had personality and inspired an industry of add-on parts and fervent owner's clubs. Ford's Ka, a fun car for the low end of the market, with innovations in style and engineering, was introduced as a concept car in 1994 at the Geneva auto show, where it was overshadowed by the New Beetle. Put on sale in 1997, the Ka captured much of the original Bug ideal: it was inexpensive, agile, compact, well engineered, and light.

The work of the young designer Chris Clements, the Ka was based on the small Ford Fiesta, but the number of parts was reduced from 3,000 to about 1,200. It was produced in a factory that relied on subcontractors, much like the one championed by Ignacio López for VW. The factory, in Valencia, Spain, became a factory park, where the makers of instrument panels or seats were located a few feet from the main assembly plant.

"Ka" came from the Egyptian word for spirit or vitality. The car's body was a complex geometry of intersecting arcs that corresponded to the juncture of its parts. Inside the Ka, as in the old Beetle, one could see straight through to the shell of the car. Its works were visible, accessible. The wheels were pushed out to the extreme corners of the body, beneath rubberized fenders that suggested the toes of dancing or climbing shoes.

From Volkswagen itself came the Lupo, smaller than the New Beetle or Golf, an inexpensive, rational, fuel-efficient car. With a new turbo direct-injection diesel engine, it achieved up to 92 miles per gallon.

Of all the new ideas, of all the claimants to the title "the real new Beetle," the most novel and potentially influential one belonged to the little pod-like vehicle that came to be called the Smart car. It was the inspiration of Nicholas Hayek, who is venerated in Switzerland as the man who revitalized the Swiss watch industry, which had been battered by Japanese competition, by developing the Swatch. As Henry Ford had taken the pocket watch as his model, the Smart car took the Swatch.

In the early 1980s, Hayek and his firm Hayek Engineering AG figured out how to replace the expensive, labor-intensive mechanical works of watches with electronic ones. A simple common mechanism—the universal appliance, taking just twelve assembly steps—was packaged in a wide variety of case designs reflecting current fashion. The watches were also issued in limited editions to create an exclusivity that appealed to collectors.

Between 1982 and 1992, the twelve-step system turned out 100 million watches in a dizzying array of colors, materials, and styles. Top designers and even artists all wanted a chance to do a Swatch; it was a prestige commission. Avid collector groups were formed, with newsletters and meetings. Today, SMK's factory in Biel, Switzerland, is a shrine. Visitors tour the factory and museum; they can even buy posters suitable for framing showing the twelve steps of Swatch assembly.

Hayek was dealing not only with the question of whether Switzerland could continue as a watchmaker but whether the modern mass production system could function economically in high-wage Western countries or would inevitably migrate to areas where labor costs were lower. Swatchism in the 1990s came to be seen as the natural successor to mass production and Fordism. In fact, it was Sloanism taken to the next level: a common technical core with infinitely variable shells.

The watch as universal appliance—the dollar pocket watch of 1900, the popularly priced wristwatch of 1950—was now succeeded by a universal mechanism with a custom exterior. Swatchism had turned the watch from jewelry into costume jewelry, driven by fashion and graphic design.

Hayek did not just create a low-end popular product, but re-created the whole Swiss watch industry by supplying movements and components for such premium brands as Omega, Rado, Longines, Tissot, and others. This pattern of customizing a universal mechanism would extend in the 1990s to beepers, cell phones, computers, and personal digital assistants. The makers of video games seized on a similar design strategy, as did the design team at Apple Computer when it offered the iMac computer in translucent colors. Cell phones were available with a seemingly endless variety of cover plates, from the American flag to Mickey Mouse to camouflage patterns. "Get a new face" said the cell phone ads.

Hayek dreamed of applying similar ideas to the automobile. To compete with the Japanese, he wanted to build clean, small, inexpensive cars. In 1991 Hayek's SMK and Volkswagen formed a partnership, with each company contributing about $10 million to develop what was soon dubbed "the Swatchmobile," but was later named the Smart car.

It was to be a small urban car with an advanced "green" power plant. "It will carry two people and two cases of beer," Hayek promised in an oft-quoted sentence. Like the Beetle it aimed to radically rethink what a car meant. "Reduce to the max" was Smart's slogan.

Many laughed at the idea of applying the Swatch model to a car. Among the skeptics was Ferdinand Piëch. When he took over at VW in January 1993, he canceled the Swatch agreement.

Hayek found another partner: Daimler-Benz. There, too, designers and engineers were skeptical. Bruno Sacco, the company's design chief, asked disdainfully, "What is a Swatch? It is something you wear on your wrist. A car?"

Sacco assigned the project to a young designer named Gerhard Steinle, who had been sent to the U.S. to establish the Mercedes satellite studio in California.

In the spring of 1994, Steinle's Smart cars were unveiled. Called the Ecospeedster and Ecosprinter, their common shape was almost that of a pod, with a visible skeleton to which body panels were attached. Using modular construction, the final assembly time was cut to four hours. Smart also pushed new modes of selling. The company began building 100 "Smart Centers" with "Smart Towers," glass showcases, like giant vending machines, to hold the vehicles. The company boasted that one could be retrieved in three minutes for a customer to test-drive.

But the car-watch analogy could work in other ways. In the prospectus for a customized Lexus show car, Giorgio Giugiaro of Italdesign, the designer of the first Golf/Rabbit, declared that "a watch is not valued by its size, but by the intrinsic value of the material it is made of, the meticulous mechanical and jewelry work involved in its making, the level of craftsmanship. This explains the difference that exists between the Swatch and a Cartier even though they do perform the same function." The difference also lay in the brands and what they implied.

Volkswagen, meanwhile, had one other retro concept to offer. The New Beetle demanded to be matched with a New Bus, but it took years of tries in the studios in Germany and California and many false starts before a concept was shown to the public. Unveiled in Detroit in 2001, the New

Microbus did not recapitulate the shapes of the old models but aimed more high-mindedly to revive the shared communal feeling of the original the counterculture loved.

"It's an idea about space, useful space, active space," the designers proclaimed about the cool techno van they had produced, sounding as if all the talk about the New Beetle had gone to their heads. They concluded deliriously that the vehicle possessed "that most elusive of qualities, a soul that was in perfect harmony with the people that drove it."

fter introducing the car in Atlanta, Dr. Piëch went to Disney
World, where Volkswagen threw another grand party to intro-
duce the New Beetle to the dealers. It was an appropriate venue.
A hotel would soon be dedicated there to Herbie the Love Bug. The same
pitch was given to VW salesmen from around the country. They left the
world of Mickey Mouse fired up to sell the New Beetle, which had been
inspired by Mickey's shape.

Other VW executives in Atlanta had visited the shrine of another
brand. "The World of Coca-Cola" was part temple to its legend, part repository of relics from its history, and part theme park, whose architecture
included a huge multistory bottle.

The Coke bottle was often said to be the most recognizable shape in the
world, the epitome of the single universal product. The Bug aspired to equal
universality. "Two shapes known around the world," one of VW's ads had
proclaimed, showing the bottle and the Bug. Andy Warhol, who reproduced
the Coke bottle in paintings, silkscreens, and prints as he did the Bug, replicating the replicated object into archetypalism, had famously said, "You
know that the President drinks Coke, Liz Taylor drinks Coke, and just think,
you can drink Coke, too. A Coke is a Coke, and no amount of money can get
you a better Coke. All the Cokes are the same and all the Cokes are good."

But when everyone can get the same thing, how are you different? Same drink, same clothes, same car—what does that do to a person's sense of individuality? And more to the point, what good does the millionaire's money do him if the beverage he buys is no better than anyone else's? The universal automobile faced the same problem, one VW hoped the New Beetle would help solve.

One Volkswagen executive from Germany ventured further into the United States. Otto Wachs went to Las Vegas. Just as Porsche, sixty years before, had visited America to learn manufacturing, Wachs had come to learn marketing—the service-driven, brand-obsessed, entertainment-inspired marketing that was as quintessentially American as the assembly line.

Wachs had served as public relations chief during the difficult months when Piëch had wooed parts boss Ignacio López from General Motors. Wachs helped Piëch keep his job despite the charges and had won the boss's trust. Now he was charged with creating a theme park for Volkswagen in Wolfsburg, to be called Autostadt, or Auto City.

Wachs visited the Las Vegas Strip where there was another "World of Coca-Cola" pavilion. An M&M candy showcase with oversized dancing ovoids was next door and nearby was Harley-Davidson, where a giant motorcycle seemed to crash through its facade. But Wachs was most amazed by Niketown, the company store decorated with relics—shoes worn by famous athletes—and equipped with elaborate multimedia displays. From time to time, the lights would dim and huge movie screens would fill with noisy, inspirational advertising videos that would have made Leni Riefenstahl proud.

This sort of spectacle disturbed Nike's many critics. They associated the hype with the arrogance of a company selling $125 shoes in poor inner city neighborhoods and its use of low-cost Asian labor. (One devastating television segment on Nike factories showed women workers sewing Nike shoes, then panned down to reveal that they were barefoot.) The critics termed the company's ubiquitous swoosh logo, the swooshstika.

Nike called its efforts "romancing the brand" and it was what Otto Wachs wanted to do at VW. Nike's marketing chiefs had long perceived the similarity between automobiles and shoes. Both products are forms of transportation in which style is a vital part of making the sale.

One day a Nike marketing manager was sitting at a stop light and glanced over at a Wolfsburg special edition Volkswagen, one of the end-of-year models dealers used to clear out inventory. The Nike manager had a brainstorm: why not special edition Nikes? Soon Foot Locker was selling Nike shoes in different colors and materials and labeled "special edition."

Nike made dozens of different kinds of shoes. Its approach was the opposite of Converse, which focused on one, the classic All Star Mays and Thomas had included in the video shown at the introduction of the Concept One.

Dating to the 1920s, the Converse All Star was another simple, classic, timeless American product, the Model T of basketball shoes, by contrast with later shoes from Converse and Adidas whose ads boasted they were "limousines for the feet."

Athletic shoes had long taken inspiration from cars. Many Nike models were sewn like auto upholstery, others were decorated with strips of material like chrome. The shoes ministered to the same fantasy 1950s cars had—that you could fly. "Implied performance" Nike called it.

Tinker Hatfield was a former track star and architect who designed most of the Air Jordan shoes, the Cadillacs of the Nike line, and frequently borrowed automobile imagery for each season's new models. Toes and heels were identified in shoe catalogues as "grilles" and "spoilers." For a shoe named for Bo Jackson, the versatile football and baseball player, he took as his model Bennie the talking taxi cab in the film *Who Framed Roger Rabbit.**

The Nike idea that Wachs most seized on was the famous line of Nike founder and chief Phil Knight that "the shoe is a vehicle for the brand." In this view, any product was no more important than the specimen of an insect or animal—it was the species that mattered. The brand was the selfish meme par excellence. No Air Jordan cross-trainer or jogging shoe existed except to further the life of the swoosh that it bore.

* In 2000, Adidas would go so far as to hire the Audi design studio in Simi Valley to design a basketball shoe for Los Angeles Lakers' star Kobe Bryant.

Did this idea apply to the New Beetle? Was the Bug now no more than a carrier for the company logo, a vehicle for conveying the brand?

The popularity of the New Beetle made it clear to VW executives that, like Mickey and Nike, the Bug was a powerful character, "evergreen" as the marketing experts called it. Didn't such a product, such a logo, such a brand deserve its own temple, like Niketown, its own theme park, like Mickey?

After Wachs's trip, Volkswagen developed a theme park, sixty-two acres of pavilions and towers, where buyers could pick up cars they had bought at local dealers. It would be called Autostadt, and stand as a testament to the way the company had changed: how VW had discovered the cult of the brand.

For his new task, Wachs delved into the world of the prophets of the New Economy and their high-minded ambitions to make companies spiritual forces. A car was more than a mode of transportation or even of personal expression. It was, in the words of one prophet, "an icon around which to enact."

Autostadt opened in June 2000 in conjunction with the Expo 2000 Festivities in Hannover. Rising beside the factory in Wolfsburg, it was part Niketown, part Legoland. It resembled a miniature world's fair ground, with gardens and water parks, an auto museum, a Ritz-Carlton hotel, and pavilions devoted to the many brands that had come under Volkswagen's corporate umbrella during the tenure of Ferdinand Piëch. More than a million people were expected to visit during the first year; the actual figure was half again as large. Autostadt was also intended to bring excitement to Wolfsburg, still widely seen as a cold, isolated city. In the 1980s, a VW designer declined a promotion that would have forced him to move there after his wife said she wouldn't live in Stylinggrad. Piëch was even said to have considered moving VW's offices out of the city.

A visit to Autostadt was meant to be a family outing. "The car is like a member of the family," Wachs said, echoing the ad pitches of the 1950s when the car was described as "a member of the family that just happens to live in the garage." Autostadt made a sacrament of the transfer of the car from the company to the owner. Dr. Ley had planned for buyers of the KdF to come to KdF-Stadt to pick up their Bugs, and it remains a tradition in Germany to pick up the car at the factory. (Sixty percent of

Porsches are picked up in Zuffenhausen, and Mercedes buyers take direct delivery of some 200,000 vehicles at three "customer centers" in Germany. Indeed, in the 1950s and 1960s, when the dollar was all powerful, American buyers would come to Wolfsburg to pick up their cars and with the money saved enjoy a two-week trip through Europe.)

The New Beetle helped sell Wolfsburg on the emotional car. It helped convince VW executives that as carmakers approach one another in quality and manufacturing costs, design and marketing will make the difference. It is an American lesson.

"Autostadt is a way to show the soul and spirit of the company," Wachs proclaimed. It was a token that Volkswagen "wants to become a company driven by emotion and culture."

Autostadt also suggested how much automobile companies had taken on the manners of show business. "There's no business without show business" proclaimed Michael Wolf, an analyst with Booz Allen consultants and author of *The Entertainment Economy*.

Toyota built an amusement park in Japan, Saturn welcomed buyers to its Tennessee factory, and both Ford and Chrysler opened small museums in Detroit.

The automotive world was abuzz with a new kind of talk, beyond just-in-time delivery, or total quality management and far beyond mass production or platform engineering. The key words of the new language were "brand," used as often as a verb as a noun, and "aspirational." An aspirational car was one that people wanted, not just needed.

The high point of the visit to Autostadt was the moment when the customer formally took delivery of the new vehicle awaiting his arrival in one of two glass towers, each twenty stories high and packed with 400 cars. The towers recalled the ones the Smart car company had erected in European cities.

The towers were the visual keynote of Autostadt, echoing the smoke-stacks of the factory's power station. (There is room for four more towers, making for a visual pun on a six-cylinder auto engine.) Mechan-ical arms move up, grab a car, then lower it to the base of the tower. Cars are fed through an underground tunnel to the customer center and their waiting buyers. The mechanism resembles a gigantic vending machine. It is a grandiose version of the clear tubes that carry pairs of Nikes from the storerooms to the floor at Niketown.

This mechanism was only one of the elements that could make it easy to laugh at VW's efforts at show biz. "To watch Germans do mar-keting," one commentator said of the place, "is like watching elephants dance: even if done clumsily it is fascinating to watch."

Although they can come by car or boat, most visitors to Autostadt arrive by train, after an hour's trip from Berlin or Hannover on the German high-speed Inter-city Express, or ICE. At the railroad station, they mount a brightly painted footbridge that crosses the Mittelland Canal, which once brought steel and coal to the factory, and restates the "Koller axis" that Peter Koller laid out as the armature for the city under Albert Speer.

They enter the Piazza, a high, glassed-in space with cafés and dis-plays. The 360-degree films and computer displays include a "Low-Impact Exhibit" where visitors take the place of crash dummies, strap-ping themselves into a vehicle that rolls down an inclined track into a barrier, then watch the results in a slow motion video. Children clamber through a forty-foot-high glass engine while their parents drive through a virtual reality trip down the Pacific Coast Highway.

But Volkswagen's vision of itself is best rendered in the museum, the work of architect Gunter Henn. Officially called the Zeithaus—literally "House of Time"—it is composed of two joined buildings. One is a rec-tangular, five-story glass "bookshelf," or "rack" as Henn calls it, holding eighty vintage cars—not just VWs but Cadillacs and Mercedeses as well—stuck in nooks like model cars in a vitrine. The other building is a softer-shaped aluminum structure housing interpretive exhibits Henn called "Memory Lane." Henn planned the two halves to juxtapose the descrip-tive and the analytic, or as he preferred it, the digital and the analog.

History is brightened and abbreviated at Autostadt. A 1936 VW3 prototype—*des Käfers Kern,* the ur-Beetle—has been painted red like a sports car. The museum boasts that it displays "the only Beetle convertible built in 1938." If so, it is the convertible in which Ferry Porsche drove Hitler from the 1938 groundbreaking ceremony for the factory.

And the present is dissected in strange ways. Inside the VW pavilions is a disturbing display of a New Beetle sliced in half through its steering wheel, the halves mechanically separating and reclosing, as if repeatedly reenacting the fate of a cartoon character split by a buzz saw. This is a trick often employed in auto show exhibits, but at Autostadt it looked anatomical, the steel and insulation suggesting the skin and fat of a dissected specimen on a table. Perhaps the New Beetle's cuteness made it seem weak and vulnerable.

Outside the museum stood the buildings dedicated to each of the brands in the Volkswagen group, alternately called "pavilions," "embassies," or "temples" by the planners.

The idea was high concept, Wachs declared: "to show the brand without showing a car," instead representing them by architecture. The four joined rings of the Audi logo are represented by four concentric rings. For Volkswagen itself, the pavilion features simple geometric shapes, cube and sphere, to suggest timeless verities: quality, safety, value. The pavilion for Bentley, the luxury brand, is faced in glass and the same green granite as the luxurious Ritz-Carlton hotel. Inside, its focus is a twenty-meter-high engine valve, symbolic of the traditional high power of the cars. Škoda's pavilion, vaguely cubist, is "imbued with a poetry which is embedded in the sociocultural context of Bohemia." For Seat the building represents Spanish "joie de vivre," a "rambla [festival]," said Henn. He added that the shape carried overtones of Frank Gehry's Guggenheim Museum in Bilbao, Spain.

These were cultural cartoons like those of Disney parks, where the French are recognized by their berets and the Germans by the lederhosen—it's a small (minded) world after all.

After Lamborghini fell under the sway of the VW empire in 1998, it was also assigned a pavilion. An executive of another marque in the VW group recalled how executives from Lamborghini were called to a meeting and asked to define the "core values" of their brand. The Italians showed impatience with the whole process. After much hesitation and head scratching, one blurted out, "Sex. Is sex."

The result is a dark cube expressing what the architects call "pure emotion" and "impetuous power." Visitors follow a twisting passage through the building. Distant music segues into the amplified heartbeat of a bull—the totem of the car—then modulates into the roar of a powerful twelve-cylinder engine and makes the whole cube shake. Visitors finally arrive in a room where a Diablo sports car hangs on the wall, glowing unearthly red, its engine in full cry.

Autostadt seemed an expanded, more serious version of a Niketown store.

Nike's architect Gordon Thompson designed the stores "to make the retail exchange of money for goods more inspiring." Thompson was inspired by the great world's fairs, notably the 1939 one in New York.

Both Niketown and Autostadt combined history—whether Jordan and Pippin relics or vintage Bugattis and Bentleys—with brand honing. Niketown, like Autostadt, was about deploying all the subbrands of the mother corporation in miniature environments, from Air Jordan to All Conditions Gear, Tiger Woods golf equipment to Mia Hamm soccer shoes. Inside each Niketown, rooms play the role the pavilions play at Autostadt.

But Niketown was a wholly more playful place. Thompson describes the shoe tube delivery system as "totally a Spacely Sprocket product." At Autostadt, Volkswagen seems to have seriously decided to be playful, rationally decided to be emotional.

Wachs, who used the word "theme" as a verb, looked outside the company for talent to design Autostadt, to professional restaurant and exhibition designers. The French decorator Andrée Putman was hired to design the interiors at the Ritz-Carlton, whose curving shape fronts the old factory. A longtime admirer of industrial architecture, she used brick to echo the brick factory buildings next door, which she compared admiringly to "a woman unaware of her own natural beauty." The smokestacks of the power plant are echoed in glass columns in the hotel lobby.

Wachs insisted that Autostadt "looks forward, it is not a monument," but it is hard not to see in the place an implicit celebration of the Piëch years. His presence is a constant reminder of how dangerous it is for VW to delve into symbolism and history.

The ground on which the new Autostadt stands, most recently a coal and oil storage area, sits beside a notch in the Mittelland Canal, a miniature harbor next to the power plant where barges brought coal and steel. It was easy to match up the notch with the World War II reconnaissance photos that showed POW, slave labor, and concentration camps around the factory and labeled the site as a "hutted encampment" with barracks for some of the 15,000 forced laborers. One of the concentration camps stood just to the north.

While Autostadt's architecture aims to speak of a bright future by spotlighting the symbolism of "brand heritage," its location had inevitably stirred the ghosts of Volkswagen's wartime past.

Over the years slave laborers had sought compensation from VW and other German companies. But courts ruled that the companies were protected from claims for compensation to slave laborers by the London Debt Settlement Agreement of 1953, which postponed the issue until the end of the reparation period for Germany. That period did not formally end until 1990, with the treaties that unified Germany. The way was laid open to a series of class-action suits filed in the U.S. as well as in Europe.

Arguing that no agreement on payments to slave laborers or statute of limitations precluded seeking damages for genocide, a group of ostarbeiter sued VW America in 1998, in New Jersey. Another suit was filed in Wisconsin by seventy-nine-year-old Anna Snopczyk, whose two-month-old child died of gross neglect in the Kinderheim in 1945.

In 1998, VW established its own $12 million fund to compensate forced laborers, seeking to address the claims without admitting legal responsibility. Dealing with war claims became an issue in the 1998 election when Chancellor Helmut Kohl declared that since the government paid out some 60 billion deutsche marks (or about $20 billion) in various claims since 1945, the rest would now be left to private industry. Kohl's opponent, the eventual winner of the race, was Gerhard Schröder, who as governor of Lower Saxony sat on the VW board. He supported a settlement for claims with equal participation by the federal government and the companies involved. Finally, after Schröder had become Chancellor, on July 17, 2000, long and tortured negotiations among governments, corporations, and international victims' rights groups ended with a formal agreement committing both companies and the government to contribute equally to a $5 billion fund for the compensation of World War II slave laborers. The agreement made moot the existing suits. However, the agreement specified that the corporate contributions were voluntary, from "moral" reasons rather than legal obligations—the same position VW had taken earlier in establishing its own fund. The companies continued to argue that under earlier agreements, the German government had assumed all reparations responsibility. And the agreement protected the companies from individual claims from plaintiffs in the U.S. or Europe.

There were estimated to be about a million and a half slave laborers alive at the end of the century, but time was on the side of the corporations; more laborers died with each passing day of negotiations.

In 1991 in a bunker inside the factory, Volkswagen established a monument to the workers who suffered at the hands of those who ran the factory in the 1940s, whom Wachs described as "those criminals" and who included the father of his boss, Piëch.

In 1992, at the end of his tenure as VW's CEO, Carl Hahn commissioned Hans Mommsen, one of Germany's most important historians, to write a history of the company during the Third Reich to help it come to terms with its history. Hahn's years in America may have made him more conscious of the company's need to come to terms with the Nazi past.

Mommsen came from a distinguished family of historians, including his twin brother, Wolfgang, his father, Wilhelm, and great-grandfather, Theodor Mommsen, who won the Nobel Prize for Literature. He and his young protégé Manfred Grieger had a budget of some $2 million, a large staff, and access to the huge VW archives. Porsche, however, would not allow access to its corporate archives, which contained much of the material on the wartime roles of Ferdinand and Ferry Porsche.

Dr. Mommsen often traveled to Wolfsburg on the InterCity Express, the ICE (the term was apt and no accident), the cool technological showpiece of the German railways. The ICE trains were named after German historical figures, from musicians to generals, from Frederick the Great to Count von Stauffenberg, the leader of the failed plot to assassinate Hitler, and, sometimes, Dr. Mommsen would smile to find himself on the train named after his great-grandfather. It was a constant reminder of the responsibilities imposed by the past, the losses and failings and the continuing possibility of retrieval.

But by the time Mommsen's book, *Das Volkswagenwerk und seine Arbeiter im Dritten Reich* (*The Volkswagen Factory and Its Workers During the Third Reich*), was published in 1996, Piëch had succeeded Hahn as CEO. Although the treatment of his father is fairly gentle, the younger Piëch was not pleased. He has so far refused to provide funds Dr. Mommsen said VW had promised to translate the book into English. "He is the kind of man who sees conspiracies," Mommsen said.

The past is inescapable at the factory, where Nazi era plaques of heroized labor still stand at the entrances. Although greatly expanded to the north, the original factory is essentially unchanged. Its most dramatic feature, the powerhouse, close to Autostadt, and the twenty-one entrance towers remain much as they were in 1940, marching along the canal in a formation as martial as any of Albert Speer's designs for Nuremberg or Berlin. One of the conical bomb shelters that dotted the facility during the war survives on the outskirts of town.

The planners tried to link Autostadt visually and symbolically back to the city of Wolfsburg. The new bridge that leads to it from the railroad station reasserts the original main axis of the town from the castle to a "crown" of civic buildings on a hill, Parthenon-style, and links the Alvar Aalto civic building and Hans Scharoun theater built in the 1960s.

Peter Koller had designed Wolfsburg's main roads wide enough to accommodate parades, but its main street, Porschestrasse, is now a pedestrian mall. Autostadt implicitly calls on the visitor to consider every detail of the environment. Autostadt is generously planted with trees and near each pavilion an appropriate type of tree has been planted. Bentley has an English oak, Škoda a Bohemian lime, VW a "cheerful yet tough and pioneering" birch, and Lamborghini a sweet chestnut, the "most Mediterranean" type tree that can survive in Wolfsburg's climate.

Even a feng shui expert was consulted. Perhaps with a nod to the dark legacies, the expert advised planners to "ground" the theme park from negative energies by planting oak and weeping willow trees at its entrance, one with powerful roots connecting it to the earth, the other with light limbs embracing the air. Autostadt managers followed the advice but renamed the willow. In German the weeping willow is called *die Trauerweide*—literally "sorrow willow." But at Autostadt it will simply be called a willow. "Sorrow," said a VW publicist, "is not a word we use at Autostadt."

Dr. Gunter Henn, the architect, saw perhaps even more significance in Autostadt's role than Wachs. He said something strange: it was necessary for places like Autostadt to exist in order to inculcate social values. Once, churches had done that, but now church and state had withdrawn from the public realm and in the post-ideological age it was left to business to stand for certain values—not just company values, but wider values—safety and concern for the environment.

Autostadt was only a first step in the process. Henn was more enthusiastic about the Crystal Factory he was also designing entirely of glass for VW in Dresden. With every part of the process visible, it was a factory designed for spectators. Customers would watch their own cars being built. It would be like looking into the shop of a master craftsman, Henn said, with sycamore wood floors and tools neatly in place. It would be like viewing artists at work—that old image of old-world craftsmanship. A German observer commented that one of the films in the Zeithaus, about the wheel, took on religious overtones. The film depicted the invention of the wheel as a kind of gift from heaven. "Religious theatricality," the German called it.

Dr. Piëch, said Henn, shamelessly flattering his patron, "was a great builder—comparable to August the Strong or Napoleon." Those men were empire builders and Piëch, acquiring his empire of brands, seemed to be doing something similar. Would he, too, overreach?

The D1 to be built in the Crystal Factory would be not a people's car but a luxury model with an innovative W-12 engine, aimed to compete with the BMW 7 series and Mercedes S class and VW's own Audis. The experience of visiting the factory to see one's car built would lend the car a sense of value that it took competitors decades to establish.

The choice of the location was purposeful. Dresden, the center of the most refined of German culture, of music and painting, opera and fine porcelain, was firebombed during World War II.

The factory, Henn explained, would be fairly small and located downtown: parts would arrive from outlying warehouses by the city's own tram system. It would have little impact on the city's infrastructure—no gritty industrial fallout. It would be high tech—tracking individual cars would be done through sophisticated information technologies. It would also mark the first automotive achievement of something like mass customization: a vision of a personal car very far from the original people's car.

The past is preserved unevenly at Autostadt and in Wolfsburg itself. Tensions are palpable in the town among longtime residents, foreign-born workers, and former East Germans. A visitor who toured Autostadt while it was under construction found a large monument to the heroes of the First World War set almost unnoticed near the highway north of the park. It was still generously covered with flowers from Armistice Day, several months before. On a gray, cold Sunday morning, wind blew garbage through downtown Wolfsburg where only a few old men strolled. Someone had covered the bust of Porsche in the town center with a blue plastic bag and sprayed it with graffiti and swastikas.

chapter twenty

I
f the Wolfsburg plant remained the company's largest and the largest in the world—"the size of Monaco" as the guides liked to say—much of the new Volkswagen's production was outside Germany. But no pavilions at Autostadt were dedicated to the company's more distant branches—in Brazil, South Africa, and China—or in Mexico and Brazil, the only ones still building the original Beetle. (Brazil stopped producing it in 1986, began again in 1992, then stopped in 1996.)

The car had become such a global product, however, that its place of assembly was almost incidental. The proportion of an automobile's price charged to its manufacture was decreasing—by the end of the century it was only 20 percent. The largest part of the cost lay in engineering, designing, and marketing.

In the 1950s, American buyers ordered cars from long option sheets, then patiently waited while they were assembled. ("Let us build you one," General Motors's advertisements urged.) But at the end of the twentieth century most buyers purchased cars off the lot and they had less sense of where the cars came from than they did when images of factories like River Rouge filled newspapers and newsreels.

Japanese cars were built in Tennessee and Kentucky, Chryslers in Canada, Fords and Chevrolets in Mexico. BMWs were built in South

Carolina, where some workers wore shirts that translated "BMW" as "Bubba Makes Wheels." Few owners of New Beetles were aware that the car was built not in Germany but in Puebla, Mexico, about a hundred miles southeast of the capital. The dispersal of automobile manufacturing around the world followed the pattern for clothing, appliances, and other products that increased the distance between consumers and factories. (There is a Nike factory not far from the VW plant in Puebla.)

Highway 150 from Mexico City to Puebla is a horizontal bazaar selling food and toys, equipment and clothing. One stretch is lined with auto repair shops with names such as El Teddy's and Willys. The signs read "Refaccionara"—repairs—"Nuevas y Usadas"—new and used. Mufflers are depicted on signs like surrealist sausages and huge tires painted with the word "Vulcanides" advertise shops that fix flats.

The logos of car manufacturers are naively drawn on the white-washed cinder block shop walls, along with childlike paintings of cars, some sad, some smiling, like cartoons. There are caricatures of Bugs that suggest a hybrid between old Beetle and New Beetle.

At a point east of the capital the road divides into a wide modern toll superhighway and a free two-lane road. It is known by several names: from the south it is the Route of Cortéz and the Conquistadors, the course they took from the ocean. From the north it is "the road of resistance to foreign intervention" or Ignacio Zaragoza Highway, named for the general who defeated the French on May 5, 1862. Like most Mexican victories over the outsider, it was short-lived, but it made Cinco de Mayo a national holiday.

Before long the road begins to rise, trucks begin to labor, and battered Beetles hunker down in second gear. Lumbering buses decorated with paintings and strips of chromed brightwork and bearing the names of girls hang almost frozen on the grades. But as night falls and one moves further out of the megacity, most of the vehicles turn off into the shantytowns. Zipping along in the left lane, their left-hand blinkers flashing impatiently as if the drivers think they are on the German autobahn, come more modern Volkswagen models, silver Jettas and black Passats, the cars of VW executives and engineers heading south to the Puebla plant.

Puebla itself is a lovely old city, the most Spanish of Mexican cities, the residents boast. The Zócalo, the main square, is lined with police cars: Beetles painted June bug silver green with blue and red roof lights.

Souvenir stands and booths selling food are everywhere. There is an unquenchable impulse to decorate and arrange. Mangoes are piled like cannonballs, the samples sliced open with serrated edges to show their blood-red interior. Many of the local handcrafted souvenirs bear licensed logos of global brands. Nike swooshes are delicately embroidered on wallets. The squat shapes of Pokémon characters are carved in onyx. There is something touching about this desperate effort to make handwork keep up with the machine age. The city is dotted with the blue VW lollipop— bus stops for workers at the plant north of town who live in boxy new apartments or old stucco houses in back alleys.

The factory spreads out on a wide plain, clinging close to the ground as if afraid to rise. On clear days, two volcanoes can be seen on the horizon. Electrical pylons rise in ranks and beneath one an old man plows a lonely field behind two mules: a photojournalistic cliché. In the lawn in front, a huge Volkswagen logo has been created in gravel. A man on his knees carefully picks stray pieces of the gravel from the grass.

Some 14,000 workers are employed here, at the highest auto wages in Mexico. They earn about $15 a day; in Germany, the workers get $27 an hour, in the U.S. $20.

When the plant opened in the mid-1960s, it was the pride of Mexican industry. But quality and labor relations deteriorated in the 1990s. The powerful national labor union, effectively a branch of the long dominant PRI party, squelched an effort to establish a local, rebel union. There have been thirteen strikes in recent years. In one, a decade ago, police took over the building. The whole workforce was fired in 1992. The old union was dissolved; a company union was set up. The collapse of the peso in 1993 and the advent of NAFTA gave management the upper hand.

Under Piëch and his executives, new management took over, just-in-time Japanese-style methods were introduced, and quality was improved.

Defects per hundred cars, which at 226 per 1,000 in 1992 were about twice the industry average, have fallen to around 96.

Puebla was the only plant where VW could manufacture the New Beetle inexpensively enough to make a profit. Workers at the plant were told that its improved quality had earned it the trust and honor of producing the New Beetle.

Despite improved conditions, the workers went on strike in the summer of 1999. The union demanded a 40 percent increase, but settled for less; the union leader, Luis Fonte, declared that the union decided to accept the settlement to avoid the cost of a strike. As the only source for the New Beetle, Puebla's workers had more leverage, but they also had more to lose if another plant was added.

Actual assembly of the car is only a small part of the process of building it. The production system introduced by José López, Piëch's expert on parts and manufacturing, changed at Puebla. Just-in-time methods were copied from an efficient Seat plant in Spain. An industrial park of suppliers surrounds the factory and prepares seats, instrument panels, electrical harnesses, and other parts. These are conveyed to the main factory with antlike efficiency by streams of wheeled carts or forklifts. Puebla is a long way from the ideal of the heroic, giant factory at River Rouge or Wolfsburg, with raw material entering at one end and finished cars emerging from the other. In the new system, work is pushed down the supply chain to contractors and their subcontractors. By specializing, the maker of seats or door handles can achieve efficiencies that a giant automaker can not. The final assembly has evolved into an assemblage of modules. Wages, too, generally fall down the chain of supply.

About two thirds of the cost of the car comes from parts purchased from suppliers. Eighty percent of these parts are German, such as Hella lamps, Bosch fuel injectors and control units, and Bayer plastics. Separate buildings around the main assembly plant house Siemens, GE Plastics, Bosch. There are piles of rolled steel from the U.S. or Germany, mounds of carpet and insulation, transmissions in cargo containers.

Inside the main factory, lamps spread an even, timeless light that contrasts with the bright sunshine outside. Modern automobile factory assembly is both surprisingly complex and surprisingly simple. Parts are stamped out in great machines and welded together, and softer parts— plastics, seats, panels, tires—are installed. The shapes of individual parts seem baffling and their junctures random. The whole process recalls a jigsaw puzzle whose apparent superabundance of parts surprisingly and suddenly resolves itself into a whole.

Welding robots work behind barriers—robot corrals. They rear and strike, then dive to apply a weld in a fiery shower. Their movements are recognizable from any number of movies featuring pit vipers or dinosaurs or other monsters. But there is something snakelike about the human workers, too, as they dart into the cars to tighten a bolt with a power tool. A welder's helmet with its swing faceplate recalls an Aztec serpent helmet. Men and machine combine in a way that makes the factory look more like Diego Rivera's mural of the River Rouge factory than River Rouge ever did.

What is remarkable is how much this automobile factory resembles the automobile factories of twenty or fifty years ago. For all the robots, there is also an amazing amount of handwork in the assembly of the New Beetle as well as the original one, especially the buffing of metal parts and welding. The images are of the classic heroic assembly line: the welder's bending concentration, the shower of sparks, the steady unreeling of the metallic wire of flux from spools. The workers wear blue overalls and baseball-style caps bearing logos of American football teams— Broncos and Cowboys. Their faces wear the concentrated self-assurance of athletes.

Although the labor agreement stipulates that at least half the work will be done by hand, workers fear machines will eventually take their jobs. The robot corrals may be there as much to protect them from the workers as vice versa.

"Robots don't buy cars," the guide explains. One hears the same line in Wolfsburg, and in both places it is supposed to sound like a humanistic affirmation of the value of handwork, happily merged with self-interest. The workers are needed not only for production, but for consumption.

Although the $15 a day these workers earn is three times the average wage in Mexico, they will never be able to afford one of the 2,500

New Beetles sold annually in the country, but they might aspire to an old one, at $7,000.

One union member in the U.S. complained:

> Volkswagen, according to its television ads, would have you believe that you can recover your lost soul by buying a new Volkswagen Beetle. Hardly, and only if your soul can be found riding on the back of a sweatshop worker making as little as $10 a day.
>
> While the New Beetle may be safer than the original, it clearly has no soul. Although it sells for $20,000, the workers who build it make only about $2,000 a year and could never afford to buy one. In contrast, the old Beetle was made in Germany and sold for about $1,500 in the 1960s. The cars were affordable to the workers who built them.

The completed New Beetles depart on special rail cars. One train heads to Veracruz to meet ships for Europe and the East Coast of the U.S., another for Acapulco and ships to Asia and the West Coast. Each hull of the fifteen-ship fleet holds 1,200 cars.

But the old cars—they are called simply Beetles—mostly stay in Mexico. In Choluca, a few miles northwest of Puebla, a VW dealership lies in the shadow of one of the largest Aztec pyramids, and a clutch of original Beetles fresh from the plant, all white, sit like eggs in front of it.

The old Beetle is considered a temporary car even in Mexico, but it has become a symbol of the endurance of Mexicans themselves. A travel writer once suggested that a bronze monument be erected to the Beetle in which two men squat on their haunches, contemplating its open engine compartment.

A section of the Puebla factory has been reserved for a special display of VWs: the harlequin Benetton model, with every part painted a different color, the wide-seated model custom-built for a visit by the portly West German Chancellor Helmut Kohl. Nearby, all the parts that go into a single car are laid out on a table for the instruction of new workers and visitor. The motors used in the original Beetles assembled at Puebla are displayed, too—descendants of the one Franz Reimspiess designed more than sixty years ago, refined, fuel-injected, and unleaded.

If the visitor's gaze strays upward to the ceiling, he sees something startlingly incongruous. In the shadows above the motor hangs a small box, softly lit from within and curtained in wine-red velvet fringed with gold. Inside is a statue of the Virgin and a vase about the size of the one on the dash of a New Beetle, filled with fresh flowers.

The cultural distance between motor and Madonna is not so far as it might seem. If Roland Barthes famously compared a new Citroën of 1955 to a cathedral, could not a simpler car like the Bug resemble a shrine or icon?

The New Beetles left the factory in Puebla with their dashboard flower vases carefully packed away inside. But cynics who thought the vase was a marketing gimmick that drivers would fill with soda straws and ballpoint pens were surprised, even touched, to pass New Beetles on the streets or in parking lots whose vases were full of zinnias and daisies and roses. Whether the flowers were real or paper, their presence was testament to the owner's affection for the car—and a synecdoche for the car's power as a repository for the fleeting blossoms of individual emotion.

The capacity of the Bug to absorb new meanings seemed endless.

One warm May day in 2000, there appeared in front of the Christie's auction house in Manhattan a clay model of the New Beetle. Volkswagen's original clay model of the Concept One was being offered at auction as part of a sale called *Masterworks: 1900–2000*. Passers-by wondered if it was a real clay model—or a real New Beetle covered with clay. It was the latter: the genuine clay model, vulnerable to heat and light and likely to crack, was kept in a climate-controlled vault inside. Lars Rachen, head of the Christie's department of twentieth century decorative arts, saw it as a symbol of the 1990s movement of revolutionizing product and brand. The clay, Rachen said, "elevated the design to sculpture." Freeman Thomas was appalled that the company would consider selling such a part of its history.

The model was estimated to fetch as much as $100,000, but it did not reach the reserve or minimum price. Instead, it came into the hands of

Jorge Pardo, an artist who displayed it several months later at the Dia Center for the Arts in New York as part of a work he called simply *Project*.

"Neither exactly retro nor appropriation," Pardo's accompanying text intoned,

> the final design of this seductive revision subtly tempers the purity of geometric exactitude toward the organic and away from what is normally coded as mechanistic, rationalistic or standardized. Simple, compact, quasi biomorphic forms connoting intimacy and a relaxed, personalized ambiance are its trademarks. Built to scale, the clay prototype no longer serves a functional purpose as the original source for the mold in the design process; a singular talisman, its role has morphed into that of emblem or icon; an anomalous model, it points to the realm of the hypothetical, the conceptual, the idea.

Talisman, icon, or ideal, was this embodiment of the New Beetle any more "true" than the Shagmobile from Mike Myers's film *Austin Powers: The Spy Who Shagged Me?* The custom-built convertible with a psychedelic paint job was sold at auction.

The more variants in which a shape is rendered the more the essence of that shape is defined. So often was the New Beetle made into models and toys, reborn as plastic kiddie cars and plush toys, that its literal manufactured version seemed simply one possible embodiment of the Beetle ideal, one body into which the restless soul of Beetleness had settled.

The best embodiment of the New Beetle gestalt, however, might have been the translucent plastic models churned out by the thousands at car shows by Volkswagen itself. The company began bringing a molding machine to shows that converted plastic pellets into five-inch models in pink and blue and purple that popped from the mold still warm to the touch and seemed as soft as protoplasm.

Some model and toy makers made Beetles whose shapes fell halfway between the old car and the new, conveniently avoiding copyright issues, but also implying some common shape. They provided an implicit answer to the question, What did the shapes of the old and new Beetles have in common, what was the basic Bug gestalt?

The old Beetle continued to be better known around the world than the New. In New York City, thousands of toy versions of a wholly mythical Beetle taxicab were sold. In Zimbabwe, a maker of tin toys built a two-foot-long Beetle from Coca-Cola and fruit cans, complete with bottle cap headlights, movable doors, and steering gear.

The makers of these models were not the only ones who rebuilt the car in their heads, but everyone who owned one. The Bug had rolled through history on wheels of irony, its story tragic and comic at once, its life longer than lives of those who built and owned it. The Bug showed that an automobile or a chair or a computer in an industrial society not only could but should carry as much cultural meaning and power as a notable building, a novel, a song, or a painting.

What the original Beetle and the New Beetle had in common was the suggestion that manufactured products might become mobile entities that attached themselves to human needs—things that used the people who used them. The marketing men with their talk of "enacting around" products were not entirely wrong. People would fall in love with a product if it found a place in the movie of their lives. The objects that lasted were those that locked themselves into collective memory by playing out roles in individual lives. A people's car could become a personal car.

The Bug carried on its strange journey not just an image or an idea but an ideal—improving people's lives by giving them a car of their own. That ideal had been perverted by monstrous ideologies, exploited by marketers, blown up into grandiosity, tattered by excesses, but it remained vital and had been the motor that both made the Bug last and made it change.

chronology

1875   Ferdinand Porsche born.

1889   Adolf Hitler born.

1893   Duryea brothers produce first American car.

1902   Porsche wins fame for developing his Mixte system, combining electric and internal combustion power.

1908   Henry Ford introduces his Model T.

1912   Porsche develops an aircraft engine.

1924   Confined to Landsberg prison, Hitler reads Ford's *My Life and Work* and formulates ideas about a people's car and highways for Germany.

1926   At the German Grand Prix, Hitler meets Ferdinand Porsche.

1928   Porsche, working for Daimler, suggests the building of a people's car. After some 15 million are produced, the Model T is replaced with the Model A Ford.

1930   Porsche establishes his own independent automobile design consultancy in Stuttgart, sketches a future racing car.

1931   Porsche offers first the manufacturer Zündapp, then NSU, his designs for a small, popular car whose design foreshadows the Volkswagen's.

1932   Porsche travels to the U.S.S.R., declines an offer to become official state designer of the Soviet Union in charge of auto industry.

Commissioned by Henry Ford's son Edsel, Diego Rivera paints his murals depicting the Ford River Rouge factory at the Detroit Institute of Arts.

1933   JANUARY 30: Hitler is elected; on February 11 his first official speech as
        Chancellor is to the Berlin auto show where he calls for "motorization,"
        a people's car, and the building of highways.

        Hitler meets with Porsche in the spring and representatives of Auto
        Union to discuss race cars.

        FALL: Porsche meets with Hitler to discuss and sketch an ideal people's car.

1934   JANUARY: Porsche presents "Exposé for a People's Car," his sketch of the
        technical specifications of a people's car. At Berlin auto show, Hitler urges
        industry to produce a Volkswagen.

        MAY 12: Hitler meets Porsche at the Kaiserhof Hotel. The Führer tells
        him the car "should look like a beetle."

        JUNE: Contract between Porsche and the state-controlled automakers'
        association, the Reichsverband der Deutschen Automobilindustrie, call-
        ing for Porsche's firm to design and test a Volkswagen that the RDA will
        then evaluate. The first three VW prototypes are built and designated
        VW3s.

1935   OCTOBER: The first prototype travels under its own power through the
        streets of Stuttgart.

        DECEMBER: A prototype is shown to Hitler at his Berchtesgaden retreat.

1936   FEBRUARY 24: Porsche shows representatives of the RDA the first Volks-
        wagen prototypes.

        OCTOBER: Porsche leaves for the U.S. accompanied by Ghislaine Kaes,
        his nephew and secretary, on the SS Bremen. He visits the Empire State
        Building and other sights and watches the running of the Vanderbilt
        Cup auto race in Westbury, Long Island. On the same day in Germany,
        testing of the three prototypes begins. Covering about 30,000 miles, it
        concludes on December 22.

        SECOND HALF OF OCTOBER: Porsche visits Detroit auto factories.

        Publication of the book Aerodynamik des Kraftfahrzeuges.

1937   JANUARY: RDA evaluates the results of the testing and finds the vehicle
        worth developing further.

        FEBRUARY: Mercedes begins construction of thirty additional test models,
        the V30s. Driven by SS drivers, these are tested for more than one and a
        half million miles between October and December.

        FEBRUARY 21: At the Berlin auto show Hitler announces that produc-
        tion of the Volkswagen will begin soon.

        MAY 6: The dirigible Hindenburg crashes in New Jersey.

MAY 22: The GeZuVor, formally Gesellschaft zur Vorbereitung des Deutschen Volkswagens, the organization to build the Bug, is established.

MAY 26: Labor leaders are beaten by Ford guards outside the River Rouge factory in what becomes famous as the "Battle of the Overpass."

MAY: Porsche returns to the United States.

JULY: Porsche's Auto Union team wins the Vanderbilt Cup race on Long Island.

1938   JANUARY: Berndt Rosemeyer is killed attempting to set a world speed record on the autobahn in a Porsche-designed super car.

FEBRUARY 24: Construction begins on VW factory near Fallersleben, in Lower Saxony.

MARCH: Hitler annexes Austria.

APRIL 20: Porsche gives Hitler a detailed model of the Volkswagen for his birthday.

MAY 26: Hitler lays cornerstone of factory.

MAY: Prototypes of the cars are shown. Tensions rise over German demands on Czechoslovakia.

JULY 3: A *New York Times* article about the cornerstone-laying ceremony compares the car to a "shiny black beetle."

AUGUST: A saving system for VW buyers is announced: savers will contribute five marks per week until they have enough for a car.

SEPTEMBER: 2,400 Italian laborers are brought in to work on the factory.

OCTOBER: The coachbuilder Reutter begins construction of additional VW prototypes.

Hitler takes over Czechoslovakia, including Bohemia, where Porsche was born.

NOVEMBER: Some 4,000 German, Italian, and Dutch workers carry out construction of the factory.

Porsche's representatives travel to the U.S. to seek machines and skilled workers. Reports of Kristallnacht on November 9 and other anti-Semitic activity in Germany hurt Porsche team's recruiting efforts in the U.S.

1939   Some fifty VW39 prototypes are constructed for publicity purposes.

FEBRUARY: Hitler appears at Berlin auto show flanked by two VW models.

Germany's leading film studio, UFA, makes a movie celebrating the Volkswagen project. Press rollout for the VW.

1939 JUNE 7: Hitler visits the factory again.
(cont'd)

AUGUST: Cross-count,.ry caravan of VWs publicizes savings plan; radio broadcasts from an Alpine pass tout the new car.

SEPTEMBER: Hitler invades Poland, setting off World War II.

1940 The first slave laborers arrive at the factory, which is still only half finished. Production of the Kübelwagen military vehicle begins.

JUNE: Hitler invades Russia.

1941 First production sedans roll off the assembly line.

1942 JANUARY 20: Wannsee Conference.

JANUARY: In a letter, Hitler orders the SS to use concentration camp labor to construct foundry for the VW factory.

FEBRUARY: Concentration camp labor arrives from Neuengamme, outside Hamburg. The workers live in Arbeitsdorf camp until October while constructing foundry.

OCTOBER: Battles of Stalingrad and El Alamein mark the turning of the war against Hitler.

1943 About 15,000 slave laborers are employed at the factory housed in several barracks camps and concentration camps. They are among 10 to 12 million used in German factories. Production of V-1 missiles at the VW factory.

1944 Some 17,365 workers are employed in war work at VW.

MARCH: Porsche requests additional slave laborers.

APRIL: Allied bombers strike the Volkswagen factory. They return several times in June and July.

Most of the factory is devoted to repairing damaged aircraft, producing bombs and shells, and manufacturing V-1 cruise missiles.

SUMMER: Dispersal of machines from factory to underground storage and manufacturing sites. Porsche leaves.

1945 APRIL 9: U.S. troops arrive in KdF-Stadt; on May 25 the town is renamed Wolfsburg. Under U.S. administration, 110 vehicles manufactured from existing parts.

SUMMER: British move in to occupy the area under postwar division agreement. Production resumes under Major Ivan Hirst.

JUNE: The factory is named Wolfsburg Motor Works.

SEPTEMBER: The factory receives a contract to produce 20,000 cars for the British Control Commission of Germany, priced at 4,000 marks each.

For the year some 1,785 cars are produced.

FALL: Ferdinand Porsche and Anton Piëch are imprisoned in France.

1946   In February, work at the plant guaranteed for five years.

A thousand cars a month are produced.

OCTOBER: 10,000th car rolls off the line in a ceremony; workers complain of inadequate food and pay.

1947   Beginning of exports, with first cars going to Holland. Dutch importer Ben Pon sketches his idea for what becomes the VW bus or transporter, built on the underpinnings of the Beetle.

Production falls to under 9,000 as economy slumps.

VW explores selling company to British and Australian firms.

Marshall Plan proposed.

1948   Heinz Nordhoff takes over as chief of company.

Group of prewar KdF savers organized to seek compensation.

JUNE: Berlin Airlift.

Exports to Switzerland and Sweden begin.

FALL: Pininfarina and Porsche are independently asked to prepare redesigned bodies for the Beetle. These are never produced.

The two-seat Hebmüller convertible begins production. A fire strikes the factory; only 682 are built.

1949   JANUARY 8: First two Beetles are shipped to U.S.

German currency reformed, establishing economy on a strong base. In June, the export model Beetle is introduced with hydraulic brakes, chrome accents, and other improvements. VW accounts for 45 percent of the value of West German industrial output.

The coachmaker Karmann begins production of the convertible Beetle at its plant in Osnabrück, Germany. Some 330,281 are produced over the years, making the car the best-selling convertible of all time.

Constitution establishing the Federal Republic of Germany (West Germany) with Konrad Adenauer as Chancellor goes into effect.

SEPTEMBER 6: Allied authorities formally transfer the Wolfsburg plant to German managers.

Porsche sports car introduced under Ferry Porsche, Ferdinand's son.

1950   MARCH: Production of the Transporter, sometimes called the Combi or Bully or bus by contrast with Bug, begins. Hydraulic brakes added to Beetle.

FALL: Ferdinand Porsche visits the factory.

1951   JANUARY: Ferdinand Porsche dies.

1952   First exports to Canada.

1953   VW of Brazil opens.

          MARCH: Split window replaced with single oval-shaped one.

          Peter Keetman photographs the factory.

1954   The film *From Our Own Strength* is made.

1955   Millionth car produced.

          Establishment of U.S. company in Englewood Cliffs, New Jersey.

          Beetle gets signal lamps in place of mechanical semaphores.

1956   South African company set up.

1957   Australian company set up.

          Karmann Ghia sports car built on Beetle platform is introduced.

          OCTOBER: Soviet Union's Sputnik, first earth satellite, launched.

1958   Edsel introduced. *Life* magazine article appears headlined "Beetle Go Home."

1959   Brazilian factory opens.

          Carl Hahn sent to U.S. as manager.

          Doyle Dane Bernbach hired as VW's U.S. advertising agency.

          Chevrolet introduces the Corvair, Ford the Falcon, and Plymouth the Valiant compact cars as 1960 models.

1961   German sales of the Beetle reach their peak.

          VW reaches settlement with KdF savers association.

          Berlin Wall is erected.

          Beetle horsepower increased to 40.

1962   Beetle sales in U.S. top 200,000.

1964   VW opens factory in Puebla, Mexico.

          Ford introduces Mustang at the New York World's Fair, bills it as the personal car you design yourself.

1965   Volkswagen buys Audi.

          Ralph Nader publishes *Unsafe at Any Speed.*

          First U.S. auto safety laws go into effect.

1966   Beetle horsepower increased to 50.

1967   Fiat passes VW in European sales.

1968   Nordhoff dies; company officially gives the Type 1 the name Beetle.

1969   MARCH: The film *The Love Bug* opens.

          JULY: Woodstock.

AUGUST: Charles Manson family murders take place.

FALL: John Muir book *How to Keep Your Volkswagen Alive* is published.

1970   Peak year of U.S. sales: more than 500,000 Beetles sold.

1971   Super Beetle introduced.

1972   After crisis at Porsche, F. A. Porsche leaves to form own design firm and Ferdinand Piëch goes to Audi.

FEBRUARY 17: The Beetle passes the Model T as the most produced car ever.

1973   FALL: Mideast war sets off energy crisis.

The Thing, also called Type 181 and Samba, a beach buggy version of the Beetle, is introduced in the U.S.

1974   The Wolfsburg factory produces its last Beetle.

"Love Bug" and "Gold Bug" models introduced in U.S.

MARCH: The Golf, the successor to the Beetle, is introduced in Europe.

1975   The Golf is introduced in the U.S. as the Rabbit.

1977   Last Beetle sedan sold in U.S.

1978   JANUARY: Last Beetle manufactured in Germany, at VW's Emden factory.

VW opens factory in Westmoreland, Pennsylvania.

1979   Last Beetle convertible sold in U.S.

1981   The 20 millionth Beetle produced—in Mexico.

1982   Carl Hahn becomes chief of VW.

Wolfsburg's Hall 54 with its robots, or "iron slaves," is opened. Audi 100 introduced.

1983   Golf 2 introduced.

Mazda's Miata introduced—a retro sports car billed as "a simple honest idea."

1984   Last Beetle imported into Europe from Mexican factory.

1986   VW buys the Spanish firm Seat.

1988   VW closes the Pennsylvania plant after slow sales and poor quality.

VW introduces Corrado sports coupé.

VW U.S. marketing chief Jim Fuller dies in Pan Am Flight 103 bombing over Scotland.

1989   NOVEMBER: Berlin Wall comes down.

1990   VW takes over former Trabant factory in Mosel.
OCTOBER: Formal reunification of Germany.

1991   OCTOBER: J Mays's Audi Avus show car is displayed at Tokyo auto show.
Memorial to war workers dedicated in Wolfsburg factory.
VW's satellite design studio opens in Simi Valley, California.

1992   Poor quality and labor difficulties in Mexican factory hurt U.S. sales.

1993   JANUARY: Ferdinand Piëch becomes VW's CEO.
J Mays and Freeman Thomas develop New Beetle designs.

1994   JANUARY: Concept One introduced at Detroit auto show.

1995   New Beetle is shifted from Polo platform to Golf platform.
Sixtieth anniversary of original Beetle celebrated.
Audi TT sports car concept shown.
J Mays and Freeman Thomas leave Audi/VW and move to SHR design consultancy.
A more advanced version of the Concept One is shown at Tokyo auto show.

1996   MARCH: Production version of what is now officially called the New Beetle shown at the Geneva auto show.
Smart car from Swatch and Mercedes shown to public.
Hans Mommsen's book *Das Volkswagenwerk und seine Arbeiter im Dritten Reich* (*The Volkswagen Factory and Its Workers During the Third Reich*) is published.

1997   Arnold Communications begins innovative VW ad campaign.

1998   FEBRUARY: New Beetle introduced to the American press.
SPRING: New Beetle goes on sale in U.S.
SEPTEMBER: VW establishes fund to compensate wartime forced laborers.

1999   Turbo model New Beetle introduced.

2000   RSI model New Beetle launched in Europe.
Autostadt opens in Wolfsburg.

2001   New microbus concept displayed at Detroit auto show.

2002   Turbo Sport New Beetle model introduced.
SUMMER: An original Beetle added to design collection of the Museum of Modern Art in New York.
FALL: Convertible New Beetle introduced.

bibliography

Andrews, Edmund L. "Germany Weighs Overhaul of 'Consensus' Capitalism." *New York Times*, February 14, 2001.

"Another Kind of Transport." *The Economist*, December 23, 1995.

Banham, Reyner. *Design by Choice*. Rizzoli, 1981.

Bradsher, Keith. "Efficiency on Wheels: U.S. Auto Industry Is Catching Up with the Japanese." *New York Times*, June 16, 2000.

Burleigh, Michael. *The Third Reich: A New History*. Hill and Wang, 2000.

Campbell, Christy. *The VW Beetle: A Celebration of the Bug*. Longmeadow Press, 1990.

Carlson, Gustav. "Evolution of a Species: Will the Bug Be Better." *New York Times*, December 25, 1994.

"The Car That Built a City." *Reader's Digest*, February 1954, condensed from *Deutsche Rundschau*.

The Center for Auto Safety, Lowell Dodge et al. *Small on Safety: The Designed-in Dangers of the Volkswagen*. Grossman Publishers, 1972.

Dawkins, Richard. *The Selfish Gene*. Oxford University Press, 1979.

DeLorenzo, Matt. *The New Beetle*. MBI Publishing Company, 1998.

The Design Museum, London. *Porsche: Design Dynasty* (Museum Catalogue), 1998.

Flat 4 Project. *Vintage Volkswagens*. Chronicle Books, 1984.

Flink, James J. *The Automobile Age*. MIT Press, 1990.

Ford, Henry, in collaboration with Samuel Crowther. *My Life and Work*. Doubleday, Page, 1924.

Forde, Gerard, and James Peto. *Ferdinand Porsche: Design Dynasty, 1900–1998*. Clifford Press, 1998.

Fox, Stephen. *The Mirror Makers: A History of American Advertising and Its Creators*. Vintage Books, 1984.

von Frankenberg, Richard Alexander. *Porsche: The Man and His Cars*. Bentley, 1969.

Gartman, David. *Auto Opium: A Social History of American Automobile Design.* Routledge, 1994.

Gelderman, Carol. *Henry Ford: The Wayward Capitalist.* Dial Press, 1981.

"German Car for Masses." *New York Times,* July 3, 1938.

Goddard, Stephen B. *Getting There: The Epic Struggle Between Road and Rail in the American Century.* University of Chicago Press, 1996.

Halliday, Jean. "Volkswagen of America: Led by the Runaway Success of the New Beetle, Carmaker Again Rides High." *Advertising Age,* December 14, 1998.

Hilsenrath, Jon E. "Ford Builds a Car to Suit India's Tastes." *Wall Street Journal,* August 8, 2000.

Hitler, Adolf. *Hitler's Table Talk, 1941–1944.* Translated by Norman Cameron and R. H. Stevens. Introduction by Hugh Trevor-Roper. Enigma, 2000.

Hogan, David J. "Innocents Abroad: The VW Beetle Comes to America, 1949–1957." *Collectible Automobile,* February 1998.

Horn, Huston. "America's Romance with a Plain Jane." *Sports Illustrated,* August 1963.

Hughes, Thomas P. *The American Genesis: A Century of Invention and Technological Enthusiasm 1870–1970.* Viking, 1989.

Jaskot, Paul B. *The Architecture of Oppression: The SS, Forced Labor and the Nazi Monumental Building Economy.* Routledge, 2000.

Jockel, Nils, and Werner Lippert. *Und läuft und läuft und läuft: Käfer, New Beetle und die perfekte Form.* Prestel, 1999.

Jockel, Nils, and Wilhelm Hornbostel. *Käfer: der Erfolkswagen.* Prestel, 1997.

Keetman, Peter. *Volkswagen: A Week at the Factory.* With essays by Armin Kley, Dirk Nishen, and Rolf Sachsse. Chronicle Books, 1985.

Keller, Maryann. *Collision: GM, Toyota and Volkswagen and the Race to Own the 21st Century.* Doubleday, 1993.

Kershaw, Ian. *Hitler: 1889–1936, Hubris.* W.W. Norton, 1998.

Kimes, Beverly Rae. *The Star and the Laurel.* Mercedes-Benz of North America, 1986.

Klemperer, Viktor. *I Will Bear Witness: A Diary of the Nazi Years, 1933–1941.* Random House, 1999.

Knepper, Mike. *Corvair Affair.* Motorbooks International, 1982.

Kries, Mateo. *Automobility: Was uns bewegt.* Vitra Design Museum, 1999.

Landes, David S. *The Wealth and Poverty of Nations.* W.W. Norton, 1999.

Lewis, David L., and Laurence Foldstein (eds.). *The Automobile and American Culture.* University of Michigan Press, 1986.

Lewis, Tom. *Divided Highways: Building the Interstate Highways, Transforming American Life.* Viking, 1997.

Ludvigsen, Karl. *Battle for the Beetle.* Bentley Publishers, 2000.

McLeod, Kate. *Beetlemania.* Smithmark Publishers, 1999.

Mitchener, Brandon. "VW's Bid to Face Wartime Past Produces Damning Book Instead." *Wall Street Journal,* November 7, 1996.

Mommsen, Hans, and Manfred Grieger. *Das Volkswagenwerk und seine Arbeiter im Dritten Reich*. Econ (Düsseldorf), 1996.

Muir, John. *How to Keep Your Volkswagen Alive*. John Muir Publications, 1969.

Nelson, Walter Henry. *Small Wonder: The Amazing Story of the Volkswagen*. Little, Brown, 1967.

"The Nostalgia Boom." *Business Week*, March 23, 1998.

Ouellette, Dan. *Volkswagen Bug Book*. Angel City Press, 1999.

Packard, Vance. *The Status Seekers*. Pocket Books, 1959.

Pasi, Alessandro. *Beetle Mania*. St. Martin's Press, 2000.

Pedersen, Martin C. "Bringing Back the Beetle: VW Reclaims Its Lost Soul." *Graphis 311*, Vol. 53, September/October 1997.

Porsche: Technology and Design. The Asahi Shimbun (Museum Catalogue), 1990.

Porsche, Ferry. *We at Porsche: The Autobiography of Dr. Ing. h.c. Ferry Porsche*. Doubleday, 1976.

"Profit at Volkswagen Rose 64.8% in 1998." *New York Times*, February 24, 1999.

Roberts, Stephen H. *The House That Hitler Built*. Harper and Brothers, 1938.

Robson, Graham, and the Auto Editors of *Consumer Guide*. *Volkswagen Chronicle*. Publications International, 1996.

Rosen, Michael J. (ed.). *My Bug*. Artisan, 1999.

Sabatès, Fabien, and Jacky Morel. *Cox en Stock*. Massin, 1993.

———. *La Cox: Une Voiture en Or*. Massin.

Schumacher, E. F. *Small Is Beautiful: Economics as if People Mattered*. HarperPerennial, 1989.

Sereny, Gitta. *Albert Speer: His Battle with Truth*. Knopf, 1995.

Setright, L. J. K. *The Designers: Great Automobiles and the Men Who Made Them*. Follett Publishing Company, 1976.

Seume, Keith. *Millennium Bug*. MBI Publishing Company, 1999.

———. *VW Beetle*. Motorbooks International, 1997.

Shi, David E. *The Simple Life: Plain Living and High Thinking in American Culture*. Oxford University Press, 1985.

Shirer, William L. *The Rise and Fall of the Third Reich*. Simon and Schuster, 1960.

Shuler, Terry. *Volkswagen: Then, Now, and Forever*. Beeman Jorgensen, 1996.

Simsa, Paul. *Die Akte: VW Käfer*. Heel, 1999.

Sloan, Alfred P., Jr. *My Years with General Motors*. Macfadden-Bartell, 1965.

Speer, Albert. Translated by Richard and Clara Winston. *Spandau: The Secret Diaries*. Pocket Books, 1977.

Speer, Albert. Translated by Joachim Neugroschel. *Infiltration*. Macmillan, 1981.

Speer, Albert. Translated by Richard Winston. *Inside the Third Reich*. Collier, 1970.

Stevenson, Peter. *Driving Forces*. Bentley Publishers, 2000.

Tagliabue, John. "Ford Europe Hopes New Small Car Won't Have Small Profit Margin." *New York Times*, January 15, 1997.

Taylor, Robert R. *The Word in Stone: The Role of Architecture in the National Socialist Ideology.* University of California Press, 1994.

Thömmes, Cornelia. *Architektur in Wolfsburg.* Berlin, 1996.

"Volkswagen and Arnold Communications Pitch a Beetle with 'More Power and Less Flower.'" *New York Times,* March 13, 1998.

Willett, John. *Art and Politics in the Weimar Period.* Pantheon Books, 1978.

Wilson, Kevin A. "The Man at the Wheel." *Autoweek,* November 16, 1998.

Womack, James P., Daniel T. Jones, and Daniel Roos. *The Machine That Changed the World.* Rawson Associates, 1990.

Wood, Jonathan. *The VW Beetle.* Motor Racing Publications, 1995.

Yates, Brock. *The Decline and Fall of the American Automobile Industry.* Vintage Books, 1983.

acknowledgments

J Mays and Freeman Thomas not only reconsidered the Bug, they re-created it. Both have been helpful and thoughtful in this and in other projects over the years. Their cooperation was invaluable.

I've learned a lot from Jerry Hirschberg, Jack Telnack, Chris Bangle, and many others in the world of automotive design.

Ed Kosner at *Esquire* first charged me with regular automobile coverage nearly a decade ago and I'm grateful to Anita Leclerc, my editor. At *Esquire* too, David Granger, Scott Omelianuk, A. J. Jacobs, and Michael Frank helped. At *The New York Times,* I'm indebted to many: Jim Cobb, Michael Cannell, Barbara Graustark, Linda Lee, and Alex Ward are only a few.

Christopher Mount gave me the memorable experience of working with him at the Museum of Modern Art and, more important, wonderful friendship. Also at MOMA, thanks to Terry Reilly, Peter Reed, Paola Antonelli, and Kim Mitchell for their support.

Aaron Betsky's Icons conference at SFMOMA gave the first impetus to rethinking the Bug. Chee Pearlman was there serendipitously with her "Bugs" issue of *ID* to publish a version of the result.

Anyone who looks at the story of Volkswagen is hugely indebted to Walter Nelson's *Small Wonder* and Karl Ludvigsen's *Battle of the Beetle.* Randall Rothenberg's *Where the Suckers Moon* is a classic and his advice was important as well. Keith Seume and Terry Shuler are respected by all Volkswagen scholars and fans, and their work has been important.

Thanks to many at Volkswagen, including: Steve Keyes, Tony Foulad-pour, Karla Waterhouse, Gerd Klauss, Maria Leonhauser, Otto Wachs, Charles Ellwood. Also, Jim Coily at Gensinger Motors, Barry Shepard and Carolyn Lanz of SHR, Fran Kelly of Arnold Communications, Liz Van-dura of VW, Peter Schreyer at Audi. The great historian Hans Mommsen graciously took time to share his work.

I'm thankful for the wisdom of Jesse Alexander, Ken Gross, Robert Cumberford, Jim Hall, Jean Jennings, and Norm Mayersohn and am grateful for important leads and suggestions to: Richard Story, Teddy Mendez, Martin Pedersen, Edward Tenner, Cliff Kulwin, Joan Ockman, Ronald Ahrens, Judith Fox, Rita Sue Siegel, Andrea Codrington, Paul Goldberger, Richard Snow, Fred Allen, Carl Magnussen, Grant Larson, Jamie Kitman, Denise McLuggage, Steve Holt, Hartmut Esslinger, Paul and Anita Lienert, Chris Wilcha, Gunter Henn, Berd Polster, Julie Lasky, Gerd Steinle, Jeff Brouws, Helen Patton, Ryan Price, Bill Tonelli, Martin Pedersen, Ken Carhone.

Special thanks for the work of my bibliographer, Caroline Patton, and the vital research and calculations provided by Andrew Patton. Kathy's eye and ear were vital.

Publisher David Rosenthal has taken me in from the cold twice now. Ruth Fecych is one of the best editors in the business and kept me on the straight and narrow, often despite my native orneriness. Both her skill and her patience are appreciated. Melanie Jackson has been a supportive agent and great friend for many years.

index

Aalto, Alvar, 87, 213
Adenauer, Konrad, 85, 87, 89
Adidas, 205
aerodynamics, 32–37, 147, 156
affinity marketing, 157, 158, 188
Afrika Korps, 69, 190
air-cooled engine, 26, 36, 65, 96, 100, 123, 174
Alfa Romeo, 42
Allen, Woody, 6, 121
Ampelmann (Lamp Man), 142–43
Anger, Kenneth, 109
Anschluss, 56
Antfarm artists collective, 122
anti-Semitism, 8, 43, 47, 51, 62–63
Antonelli, Paolo, 156
Apple Computers, 132, 181, 199
Arnheim, Rudolf, 195
Arnold Communications, 187, 188
art cars, 133, 134
Art Center College of Design, 152, 153
Art Nouveau, 62
Aschwanden, Peter, 118
assembly line, 12, 40, 41, 47–48, 63, 127, 174, 221
athletic shoes, 204–6, 208, 210, 218, 219
Audi, 18, 150, 152, 158, 159, 178, 181, 209
    Quattro four-wheel drive system, 147, 160, 177
    TT sports car, 166, 167
Auschwitz, 75–76
Aus eigener Kraft (film), 88
Austin, 42, 47, 156
Austin Powers films, 3, 224
autobahns, 20–24, 29–30, 35, 42, 54, 57–58, 65, 84

Automobil Verkehrs- und Übungs-Strasse (AVUS), 15, 19–21
Auto Union, 18, 28, 42, 80, 93, 141, 160, 167
    race cars produced for Hitler by, 19, 24, 43, 57, 58, 136, 167, 186
Autostadt theme park, 204, 206–11, 213–15, 217

Baader-Meinhof gang, 183
Banham, Reyner, 107–8
Barker, Clyde, 40
Barris, George, 110
Barthes, Roland, 162, 223
Bauhaus, 33, 34, 37, 87, 95, 152, 153
Bayer Corporation, 51, 173, 220
Beach Boys, 113, 151
beatniks, 98, 103
Bennett, Harry, 43
Bentley, 178–79, 209, 210, 214
Bergen-Belsen, 75
Berlin Blockade, 83
Berlin Olympics (1936), 42
Berlin Wall, 139–41, 150, 183
Bernbach, William, 92–95
Bitomsky, Hartmut, 144
Blank, Harrod, 134
Blitzkrieg, 20
BMW, 146, 152, 159, 160, 178, 215
    bubble car produced by, 139–40
    Hitler and, 28, 50
    Isetta, 139–40
    in postwar Germany, 80, 85
    South Carolina factory of, 217–18
Boorstin, Daniel, 45
Booz Allen, 207

Bosch, 220
Bostwich, David, 195
Braun, Wernher von, 77, 80, 99
Braun electronics, 98, 153, 154
Brazil (Updike), 144n
Breech, Ernie, 82, 99
British Leyland, 130
Brodovich, Alexy, 95
Brown, Jerry, 123
Browne, Hablot, 174
Brunswick und Lüneburg, Duke Franz von,
     55
Bryant, Kobe, 205n
bubble cars, 85, 99, 139
Buchelhofer, Robert, 180
Bugatti, 178, 210
Buicks, 44, 153, 182, 197
Bullitt (film), 109, 154
Burney Streamliner, 35
Byrum, Captain Clifford, 79

Cadillac Ranch, 122
Cadillac, 11, 44, 48, 97, 102, 105, 126, 182, 208
Caen, Herb, 98
Caine, Michael, 194
Cal Look Beetle, 109, 191
California Air Resources Board (CARB),
     155–56
Campbell, Malcolm, 57
cameras, 4, 12, 98, 174
Caracciola, Rudi, 19
Carpenter, John, 111
Carter, Jimmy, 123
Castaing, François, 196
Chamberlain, Wilt, 172
Chaplin, Charlie, 12, 94, 105, 110, 175
Checkpoint Charlie Museum, 139
Chenowth dune buggies, 115–16
Chevrolet, 44, 45, 95, 151, 181–82, 197, 217
     Caprice, 157n
     Corvair, 99, 100, 126
     Monza, 100
     Vega, 124, 125
Chicago World's Fair (1893), 174
Christie's auction house, 223
Christine (film), 111–12
Chrysler Corporation, 100, 111, 157, 168, 196,
     207
     Airflow, 32, 34, 37
     Neon, 2, 157, 195
     PT Cruiser, 194–96
Churchill, Winston, 83
Cianetti, Tulio, 74

Citroën, 2, 89, 162, 223
Clements, Chris, 198
Clements, F.G.H., 21
Coca-Cola Corporation, 195, 203, 204
Cold War, 2, 83, 84, 87, 99, 139, 141, 143, 153,
     183
compact cars, 92, 99–101, 126
computers, 3, 4, 6, 132, 181, 199
concentration camps, 73–77
concept cars, 155, 161, 162
Conquest of Cool, The (Frank), 103
Continental Tire Company, 141
Converse athletic shoes, 205
Corrados, 191
counterculture, 103, 110, 116–19, 122, 123,
     132, 137, 139
Crumb, R., 118
Crystal Factory, 214–15
customization, 109–10, 191–93

Dachau, 75
Daimler-Benz, 19, 28, 29, 200
Dane, Maxwell, 93
Darwinian theory, 4
Dasher, 125
Datsun, 125
Dawkins, Richard, 4–5
Dean, James, 99
DeMille, Cecil B., 46
denazification, 79, 81
Depression, 18
Detroit Institute of Arts, 40
Deutsche Arbeitsfront (DAF; German Labor
     Front), 50–52, 54, 55, 60, 61, 63, 74, 80,
     191, 192
Deux Chevaux, 138
Dia Center for the Arts, 224
Dichter, Ernest, 101
Disney, Walt, 12, 94, 109, 110–12, 175
Disney Company, 189, 193, 209
DKW, 18, 28
Doblin, Jay, 13
Dodge, 111, 152
Doyle, Ned, 93
Doyle Dane Bernbach (DDB), 92–95, 97,
     102, 104, 141, 159, 187
Drysdale, Don, 110
Dubonnet, 35
dune buggy, 109, 114–16, 122, 189
Dürer, Albrecht, 32
Duryea brothers, 10
Dutch, Von, 110
Dymaxion Car, 33

Earl, Harley, 46, 92
Eastman, George, 12, 174
Eddy, Don, 122
Edsel, 100
Eifelrennen race, 20
Elam, Kimberly, 172
Ellwood, Craig, 152
emissions standards, 108, 155–56, 190
energy crisis, 123, 124
Exner, Virgil, 111, 136–37
Expo 2000, 206

Family, The (Sanders), 114
Fahrvergnügen ad campaign, 147
Ferdinand, Archduke, 16
Ferrari, 97, 125
Fiat, 18, 39, 47–48, 103, 130, 158
    Brava, 2
    Topolino, 2, 48, 137–38
Fonte, Louis, 220
Ford, 2, 33, 39–41, 88, 95, 115, 138, 168, 207,
        217
    Cologne plant, 8, 47
    Falcon, 99, 100
    Fiesta, 198
    Ikon, 196
    Ka, 197
    Model A, 44, 109
    Model T, 9, 10, 12–13, 39, 44–46, 88, 105,
        124, 134, 140, 170
    Mustang, 100, 109, 154
    River Rouge plant, 39, 40, 43, 44, 47, 54,
        95, 127, 217
    Taurus, 146, 147
    Thunderbird, 194
Ford, Edsel, 40
Ford, Henry, 10–12, 39, 40, 54, 65, 80, 186, 198
    autobiography of, 8, 11, 45
    mass production methods of, 10–11, 41, 48
    Nazis and, 47
    Porsche and, 43, 46, 70
Ford, Henry, II, 82
Fox, 125
Fox, Stephen, 93–94
Frank, Thomas, 103
Freud, Sigmund, 195
Friedrich, Caspar David, 65
fuel efficiency, 197, 198
Fuller, Buckminster, 33
Fuller, James, 147

Gage, Robert, 95
Ganz, Joseph, 8

Garcia, Jerry, 137
Garfield, Bob, 104, 188
Gargarin, Yuri, 99
GE Plastics, 220
Gehry, Frank, 209
General Motors, 42, 82, 126, 145, 217
    in China, 197
    Frigidaires produced at, 53–54
    López at, 179–80, 204
    Lordstown plant, 124
    marketing innovations at, 44–46, 102
    Motorama shows, 155
    Opel and, 28, 48, 179
    Saab and, 157
    unionization at, 41, 43
    See also Buick; Cadillac; Chevrolet
German Grand Prix, 15, 19
German Werkbund, 62
Gesellschaft zur Vorbereitung des
        Deutschen Volkswagens (GeZuVor), 54
GeZuVor, 54
Giacosa, Dante, 137–38
Giugiaro, Giorgio, 125, 200
Goebbels, Joseph, 50, 51, 53
Gogomobil, 85
Gold Rush of 1849, 5
Golf, 125, 158, 165, 166, 173, 178, 186, 187, 191,
        198, 200
Göring, Hermann, 51, 57, 68, 69
Gould, Stephen Jay, 175–76
Graf Zeppelin, 21
Grapes of Wrath, The (film), 196n
Grass, Günter, 6, 101
Grateful Dead, 137, 169
Graves, Michael, 151n
Great Present and the Greater Future, The
        (Ford), 8
Greenburg, Dan, 104
Grey Advertising, 93
Grieger, Manfred, 212
Gropius, Walter, 34
Gross, Elly, 76
Grundig, 181
GTIs, 147, 191
Gugelot, Hans, 153

Hahn, Carl, 93, 141–42, 145, 147–48, 159,
        178, 212, 213
Hall, James, 186
Harris, Robert, 63n
Hatfield, Tinker, 205
Hayek, Nicholas, 198–200
Hebmüller convertible, 135

Heckhausen, Marcus, 143
Heller, Steven, 62
Henn, Gunter, 208, 209, 214
Hickel, Walter, 110
highways, 10, 41–42
    See also autobahns
Himmler, Heinrich, 55, 68, 71, 72
Hindenburg disaster, 43
Hine, Thomas, 102, 131
Hirst, Major Ivan, 80–82, 84, 123
Hitler, Adolf, 2, 18, 61, 64, 86–89, 91, 93, 94,
    144n, 190
    assassination plot against, 213
    auto racing and, 15, 20, 43, 57–58
    autobahn plans of, 21–24, 42
    Beetle concept of, 31–33, 36, 170
    Berlin auto show speeches of, 7–8, 18,
        27–28, 49, 55, 84
    DAF and, 51, 52
    Ford and, 11, 41, 47, 65
    imprisonment of, 8–9, 11
    Porsche and, 15–16, 19, 25, 26, 28, 29, 44,
        50, 57, 67, 84, 136, 150
    rise to power of, 17
    swastika symbol of, 62
    and Versailles Treaty, 35, 83
    Volksempfänger program of, 52–54
    at VW factory site, 55–56, 58–60, 63, 67,
        135, 209
    during World War II, 69–71, 75
Hoffman, Max, 91
Hoffmann von Fallersleben, Heinrich, 55
Honda, 125, 131, 145, 146, 148
Hopper, Admiral Grace, 6
Horch, 18, 69
hot rods, 109
How to Keep Your Volkswagen Alive (Muir),
    116–18
Hubbell, Sue, 32
Huhnlein, Korpsführer Adolf, 20–21
Humber, 82
Huxley, Aldous, 41
hyper-cars, 197

I.G. Metall, 129–30
Iacocca, Lee, 100
IBM PC clones, 4, 132
Insanely Great (Levy), 132
Institute for Automobile Safety, 125–26
interchangeable parts, 11, 48
International Style, 102
Issigonis, Alex, 138, 194
Italdesign, 200

Italian Job, The (film), 194

Jackson, Bo, 205
Jaguar, 91
Jarry, Paul, 36
Jeep, 2, 68, 101, 175, 195
Jensen, Lance, 187
Jetta(s), 125, 158, 166, 186, 191, 219
Jobs, Steve, 4, 132, 181
Johnson, Paul, 9
Johnson, Philip, 34
Jordan, Mark, 157n
Jung, Carl, 4, 195
just-in-time production, 128, 130, 145, 178,
    219, 220

Kaes, Ghislaine, 41, 42
Kafka, Franz, 32, 41
Kales, Joseph, 17, 19, 27
Kamm, Wunibald, 36
Kapuschinski, Ryszard, 124
Karmann Ghia, 111, 136–37
Keaton, Buster, 110
Keetman, Peter, 87–88, 95, 144
Kellogg's, 195
Kennedy, Edward M., 117
Kennedy, John F., 100
Kersting, Walter Marie, 53
Kesey, Ken, 117, 137
Kinderheim, 76, 79, 211
Kirk, Claude, 110
Kita, 151n
Klein, Lilly, 76
Klemperer, Viktor, 22–23
Klemperer, Wolfgang, 36
Knight, Phil, 205
Kocks, Klaus, 180
Kodak Brownie camera, 4, 12, 174
Koenig, Julian, 92, 104
Kohl, Helmut, 142, 211–12, 222
Koller, Peter, 56, 86–87, 208, 213
Komenda, Erwin, 17, 19, 136, 172
König-Fachsenfeld, Reinhard Freiherr von,
    36
Korbel, Hans, 76, 79
Kraft durch Freude (KdF; Strength through
    Joy) , 52–53, 55, 61, 63, 65, 68–70
    KdF-Stadt, 56, 71, 80, 81, 206
    KdF-Wagen, 59, 60, 64–65, 69, 105, 112,
        190–91
Kraft Foods, 195
Kristallnacht, 47
Krone, Helmut, 92, 98

Kroner, Kurt, 96, 128
Krupps steelworks, 136n
Krups appliances, 154
Kübelwagen, 68–69, 122, 190
Kustom Kar Kommandos (film), 109
Lafferentz, Bodo, 54, 58, 59, 74
Lagaay, Harm, 154
Laird, Melvin, 110
Lamborghini, 178, 209–10, 214
Lantz, Carolyn, 162
Lawner, Ron, 187, 188
Le Corbusier, 34, 138
Ledwinka, Hans, 36, 175
Lee, Don, 46
Leica cameras, 98
Lerbs, Eberhard, 190
Levy, Steven, 132
Lexus, 200
Ley, Robert, 51–55, 58–61, 69, 74, 79–80,
        206
Lienert, Paul, 146–47
Lincoln, 42, 124, 125
    Zephyr, 37
Lindbergh, Charles, 42
Lockheed, 28, 116
Loewy, Raymond, 34–35, 54, 92, 97
Lohner company, 16
Lois, George, 94
London Debt Settlement Agreement, 11
Long Island Motor Parkway, 10, 41–42
López de Arriortúa, José Ignacio, 179–80,
        198, 204, 220
Lorenz, Konrad, 176
Lothair II, Emperor, 55
Love Bug movies, 110–13, 122, 190, 192, 203
Lovins, Amory, 197
Luftwaffe, 68, 91
Lupo, 181, 198
Lutz, Bob, 195
luxury cars, 17, 102, 181–82

Machine That Changed the World, The
        (Womack), 128
MacMurray, Fred, 110
Manson, Charles, 109, 113–15
Manson, Marilyn, 115
Marengo, Jim, 147
Mark II computer, 6
Marsh, John, 149, 150, 154, 163
Marshall, George, 83
Marx, Karl, 183
Maserati, 125
mass production, 10–12, 33, 40–41, 44, 174

See also assembly line
Mattel toy companay, 116
Mauldin, Bill, 69
Mays, J, 150–54, 167–68, 193–95, 205
    New Beetle designed by, 150–51, 156, 158,
        160, 161, 163, 164, 171, 185, 188
Mazda Miata, 156–57
McDonald, Thomas, 24
McEvoy, Colonel Michael, 81, 123
McLuhan, Marshall, 97
McNamara, Robert, 100
McQueen, Steve, 109, 154
memes, 4–5
Mencken, H. L., 6
Mendez, Ted, 192–93
Mengele, Josef, 76
Mercedes, 10, 124, 153, 159, 207, 208, 215
    Autobahn Courier, 23
    Hitler's, 9, 56, 59, 67
    and people's car, 24, 28, 29
    in postwar Germany, 80, 85
    race cars built by, 16, 19–20, 46, 57, 186
    in U.S., 91, 98, 146, 160, 200
Mercury, 109
Messerschmitt, 85, 98–99
Meyers Manx, 109
MG, 101, 156, 194
Michelin brothers, 174
Michels, Doug, 122
microbus, 117, 132, 137, 200–201
Mini, 2, 83, 125, 138–39, 194
Mitbestimmung, 129
Mixte system, 16
modernism, 33–34, 37, 87, 95, 133, 153
Moennich, Hans, 88
Mommsen, Hans, 212–13
Monk, Thelonius, 94
Moog, Otto, 39
Morris, 47
Morris, Dave, 158
Moses, Robert, 42
motivational research, 101
Mouse, land ship, 69
Muir, Eve, 117–19
Muir, John, 116–18
Museum of Modern Art, 34–35, 156
Mussolini, Benito, 57, 74
Myers, Mike, 224

Nabokov, Vladimir, 32
Nader, Ralph, 125–27
NAFTA, 219
Nash Rambler, 99

National Socialist Driver Corps, 20
National Sporting Commission for German
    Racing, 20
Nazis, 2, 8, 56, 57, 66, 67, 69, 70, 79, 82, 97,
    101, 136, 190, 212, 213
    autobahn program of, 23
    consolidation of power of, 25
    Ford and, 47
    industrial production under, 68, 73, 87
    Labor Front of, *see* Deutsche Arbeitsfront
    Motorisierung doctrine of, 7
    Porsche and, 29, 43
    propaganda of, 20, 53–54, 61
    slave labor used by, 72, 75, 76
    swastika symbol of, 62
Nesbitt, Bryan, 196
Neumann, Jens, 185
Neumeyer, Fritz, 17
Neue Sachlichkeit (New Objectivity), 88
Neurath, Otto, 62
New Beetle, 2–3, 167, 169, 181, 194, 196–98,
    200–201, 218, 225
    ad campaign for, 187–88
    auction of clay model of, 223–24
    bud vase in, 162, 169–70, 223
    customization of, 191–93
    design of, 149–51, 154–55, 158, 160–64, 168
    engineering of, 165–66
    introduction to press of, 177–78, 182–83
    limited edition versions of, 189–90
    Mexican plant manufacturing, 218,
        220–23
    pop culture and, 186–87, 203–4, 206, 207
    rub-off effect of, 185–87
    sales of, 189
    shape of, 170–74, 176
New Deal, 23
Newton, Helmut, 188
New York World's Fairs, 42, 100
Nicholson, Geoff, 135–36
Nike, 204–6, 208, 210, 218, 219
Nissan, 125, 128, 142, 148
Nordhoff, Heinrich (Heinz), 29, 49, 82,
    84–86, 93, 108, 129
Norman, Donald, 175
NSKK, 55
NSU, 17–18, 24, 25
Nuremberg War Crimes Tribunal, 79
Nuvolari, Tazio, 42

Oertzen, Klaus D. von, 18, 19
Oldenburg, Claes, 161
Oldsmobile, 44

Opel, 28, 29, 48–50, 82, 85, 142, 179
Organization for Economic Cooperation
    and Development, 84
Overy, R. J., 23

Packard, 42, 97, 102
Packard, Vance, 101, 102
Panzers, 69
Papanek, Victor, 123
Pardo, Jorge, 224
Parker, Bonnie, 40
Passat, 145, 179, 185–86, 219
Peglau, Karl, 143
Pennsylvania Turnpike, 42, 224
Persian Gulf War, 115
Peugeot, 39, 158
Piëch, Anton, 57, 74, 80, 159
Piëch, Ferdinand, 159–60, 164, 167, 178–80,
    200, 213–15, 219–20
    Autostadt and, 206, 211
    New Beetle and, 161, 165, 166, 177–78,
        180–82, 185, 203
Pininfarina, 97
Pirsig, Robert, 117
planned obsolescence, 45, 102
platforms, 165–66, 178, 181
Plymouth, 111, 99
Poe, Edgar Allan, 2
Pon, Ben, 91, 137
Pontiac, 111
Pop Art, 122, 161
Popeye comic strip, 2, 175
Popp, Franz Josef, 50
Populuxe (Hine), 131
Porsche, 99, 167, 172, 177, 180, 181, 207,
    212–13
    Cayenne sport utility vehicle, 180
    China and, 196
    racing program of, 159
    Thomas as designer for, 153, 154
Porsche, F. A., 159, 180–81
Porsche, Ferdinand, 15–19, 24–29, 36, 40,
    62, 65, 91, 96, 117, 159, 215
    and Beetle design, 31, 32
    at groundbreaking ceremony for VW
        factory, 56, 60
    Hitler and, 15–16, 19, 24–26, 28, 29, 44,
        50, 57, 67, 84, 136, 150
    postwar arrest and detention of, 80,
        99
    racing cars of, 15–19, 24, 35, 57–58, 65,
        172
    U.S. visits of, 41–44, 46, 47, 49, 95, 204

during World War II, 68–74, 213
Porsche, Ferry, 16, 43, 57, 60, 80, 136, 159,
    181, 209, 213
Porsche, Louise, 16, 159
Presley, Elvis, 109
Presley, Priscilla, 100
Potsdaflagreements, 80
Putnam, Andrée, 210
P-Wagen, 19–20

Quantum, 125

Rabbit, 125, 145, 200
Rabe, Hans, 27
Rabe, Karl, 17, 19
Rachen, Lars, 223
racing cars, 10, 15–21, 24, 35, 36, 43, 57–58,
    65, 99, 167, 172
radios, 50, 52–54, 131
Railton, Arthur, 95
Rams, Dieter, 153
Rand, Paul, 95
Rapaille, Clothaire, 195
Raubal, Geli, 9
Reagan, Ronald, 149
refrigerators, 53–54
Reichsverband der Deutschen Automobilin-
    dustrie (RDA), 28–30, 49–51
Reimspiess, Franz Xaver, 17, 27, 61, 68, 117,
    163, 223
Renault, 80, 89, 101, 103, 130, 158
    Fiftie, 194
    Twingo, 2, 197
Renger-Patzsch, Otto, 88
Restoration Hardware, 193–94
retro design, 2, 156, 193–95, 200
Reutter, 29
Riefenstahl, Leni, 182, 204
Rivera, Diego, 40, 221
Roberts, Alan, 17
Roberts, Stephen, 22–24, 51
robots, 130, 144, 221
Rodgers, Will, 155
Rolls-Royce, 124
Rommel, General Erwin, 68
Roosevelt Raceway, 42
Rootes, William, 82
Rosemeyer, Berndt, 21, 43, 58
Rosenau brothers, 190
Rosenberger, Adolf, 17
Roth, Ed "Big Daddy," 109
Rumpler Teardrop, 35
Saab, 157

Sacco, Bruno, 200
safety standards, 108, 127, 173, 189
Sale, Kirkpatrick, 123
Samsung, 181
Sanders, Ed, 114
Saturn, 207
Sauckel, Fritz, 73–76, 79
Schaefer, Herbert, 160
Scharoun, Hans, 87, 213
Schirach, Baldur von, 8
Schmidt, Ellsa, 76, 79
Schreck, Julius, 19
Schreyer, Peter, 150–51
Schröder, Gerhard, 212
Schulenberg, Count von, 54–56
Schumacher, E. F., 122–23, 197
Schwonke, Martin, 143
Scirocco(s), 191
Sears Roebuck, 54
Seat, 166, 178, 179, 209, 220
Segar, E. C., 174
Seiffert, Ulrich, 161, 164
Selassie, Haile, 123–24
Sellers, Peter, 194
Setright, L.J.K., 172, 173
Sheeler, Charles, 40, 88
Shepherd, Barry, 155, 185
Shirer, William, 63
SHR Perceptual Management, 154–55, 161,
    162, 167–68, 170, 185
Siemens, 181, 220
Silver Arrows, 20, 35, 36, 57, 58, 65, 172
Škoda, 142, 166, 177, 209, 214
slave labor, 70, 72–77, 79, 211–12
Sleeper (film), 121
Sloan, Alfred, 44–46, 48, 102
Small Is Beautiful (Schumacher), 122–23
Small on Safety (Nader), 126
Smart car, 198–200, 207
space race, 98, 99, 105
Snopczyk, Anna, 76, 21
Sony, 131
Sottsass, Ettore, 151n
Speer, Albert, 56, 70–73, 80, 208, 213
Spitfire, 82
Sputnik, 98, 99, 103
SS, 29, 37, 60, 68, 69, 71, 82
    at groundbreaking for VW factory, 55
    logo of, 63, 190
    Rosemeyer in, 43
    slave labor of, 72–74, 79
Stalin, Joseph, 83
Standard automobile company, 8, 61

Stanton, Arthur, 93
Starr, Ringo, 194
Status Seekers, The (Packard), 102
Steinle, Gerhard, 200
Stephenson, Frank, 194
Still Life with Volkswagen (Nicholson),
     135–36
Stout, William, 33
streamlining, 23, 32–35, 37, 42, 54
Street Sleeper (Nicholson), 135
Strength through Joy, *see* Kraft durch Freude
Stuck, Hans, 19–20
Studebaker, 95, 97, 102
Stuka dive bombers, 20
Stuttgart University, 36
Suez crisis, 194
swastika, 60–62, 74
Swatch, 189, 198–200
Swiss Grand Prix, 20

Taeschner, Titus, 87
tanks, 69, 98
Tatra, 36, 37, 175
Taylor, Frederick, 39
telegraph, 6
Tenner, Edward, 5–6
There Stands America (Moog), 39
Thing, the, 122, 190
Thomas, Freeman, 153–54, 166–68, 193, 205,
     223
     New Beetle designed by, 150–51, 156, 158,
     160, 163, 164, 170, 171, 185, 188
Thompson, Gordon, 210
Thyssen, Fritz, 51
Tinbergen, Jan, 175
Tin Drum, The (Grass), 6, 10
Todt, Fritz, 21, 23, 36, 73
torsion bar suspension, 17, 26, 65, 123, 138
Toyota, 125, 127, 128, 142, 145, 146, 148, 207
Trabant, 139–43
tractors, 18
Triumphs, 156
Tropicana orange juice company, 191
Trotsky, Leon, 40
Tucker, Preston, 175

U.S. Army medical corps, 79
Ulm, design school of, 153
United Auto Workers, 41, 43
Unsafe at Any Speed (Nader), 126
Updike, John, 144n

V-1 and V-2 rockets, 75, 77

Valkenhayn, Fritz von, 17–18
Van de Kamp, Will, 91
Vanderbilt, William K., Jr., 10, 41–42
Vanderbilt Cup race, 10, 41–43
Vanzura, Liz, 187, 188
Velvet Monkey Wrench, The (Muir), 118
Versailles Treaty, 35, 83
Vienna Secession, 62
Vietnam War, 110
Vintage Volkswagen Club, 133
visual positioning, 154, 155
Volksempfänger (people's radio), 50,
     52–54
Vorwig, Wilhelm, 49, 50
Vostok space capsule, 99
VW Komplex, Der (film), 144

Wachs, Otto, 180, 204–7, 209–12, 214
Walkman, 131
Walsh, Bill, 110
Wanderer Werke, 17, 18, 62
Warhol, Andy, 122, 203
Warkuss, Hartmut, 152–53, 158, 160
Warner, Joseph, 43
Warrilow, Clive, 185
Wartburg, 140
watches, 198–200
Werlin, Jacob, 9, 24, 59, 72–73
Wesselman, Tom, 122
Weston, Edward, 88
White, E. B., 13
Who Framed Roger Rabbit (film), 111, 205
Why Things Bite Back (Tenner), 5
Wiener Werkstätte, 62
Wilhite, Steve, 187, 188
Williams, Robert, 109
Wilson, Dennis, 113
Wilson, Woodrow, 10
Wolf, Michael, 207
Womack, James, 128
Woodley, Richard, 158
Woodstock, 117
World War I, 20, 27, 51, 69, 215
World War II, 63, 68–77, 79, 86, 159, 211,
     212, 215
Wozniak, Steve, 132

Zen and the Art of Motorcycle Maintenance
     (Pirsig), 117
Zeppelin works, 36
Zündapp, 17, 24

about the author

phil Patton is the author of *Dreamland: Travels Inside the Secret World of Roswell and Area 51,* and is a regular writer for the *New York Times* "Public Eye" and "Design Notebook" columns. He is a contributing editor at *Esquire, Wired,* and *ID,* and a consulting curator of the Museum of Modern Art exhibition "Different Roads: Automobiles for the New Century." He has taught at the Columbia Graduate School of Journalism and served as a commentator for CBS news, the History Channel, and several public television series. He lives in New Jersey; his Web site is at philpatton.com.